The Everyday Physics of Hearing and Vision (Second Edition)

The Everyday Physics of Hearing and Vision (Second Edition)

Benjamin de Mayo

University of West Georgia, Atlanta, GA, USA

IOP Publishing, Bristol, UK

ISBN   978-0-7503-3207-1 (ebook)
ISBN   978-0-7503-3205-7 (print)
ISBN   978-0-7503-3208-8 (myPrint)
ISBN   978-0-7503-3206-4 (mobi)

DOI   10.1088/978-0-7503-3207-1

Version: 20210101

IOP ebooks

British Library Cataloguing-in-Publication Data: A catalogue record for this book is available from the British Library.

Published by IOP Publishing, wholly owned by The Institute of Physics, London

IOP Publishing, Temple Circus, Temple Way, Bristol, BS1 6HG, UK

US Office: IOP Publishing, Inc., 190 North Independence Mall West, Suite 601, Philadelphia, PA 19106, USA

Contents

Preface

During my early days as a physics professor, I developed a class for non-science majors called 'The Physics of Music and Art'. It was very popular among education and business majors, as well as art and music students. A number of acoustics and optics books were available. Yet, none dealt with both subjects together—so I wrote one! The book that I wrote for my class told how light waves and sound waves affect us through vision and hearing. This book, 'The Everyday Physics of Hearing and Vision', is an outcome of that early book. As before, it is a non-technical treatment of the physics of how we see and hear. Lots of topics are considered, including diamonds, mirages, musical instruments, and color mixing.

The second edition expands on the earlier topics. It also includes new developments as well, such as the new hearing aids, augmented vision devices, and gene therapy for hearing and vision problems.

We live in an amazing world, and to a large extent, our links to this world are through our eyes and ears. I hope that you will find that this second edition expands your relation to this world. And I hope that you will find the book as interesting, enjoyable, and informative as I did in writing and illustrating[1] it.

---

[1] All of the figures, drawings, diagrams, and photographs are the work of the author, unless otherwise noted.

Acknowledgments

To my wonderful wife, Judy, who has been of immeasurable assistance and excellent companionship, and to many wonderful students, I dedicate this book. Also, I gratefully acknowledge the generous and crucial support of the University of West Georgia and the Georgia Space Grant Consortium – NASA. The author is grateful for the efforts of the IOP Publishing personnel, especially Editorial Assistant Sarah Armstrong and the Production staff.

Author biography

Benjamin de Mayo

 Ben de Mayo is a native of Atlanta, Georgia, USA. He attended Emory and Yale Universities and the Georgia Institute of Technology, receiving a BS, MS and PhD, all in physics. After a post-doctoral research associateship in metallurgy at the University of Illinois, de Mayo has taught at the University of West Georgia. For 24 years, de Mayo has received a small grant from Georgia Space Grant Consortium – NASA to do research with undergraduate science majors and to present science programs for the public and for school groups. De Mayo's interests include hiking, fishing, and sculpture. His research areas are high temperature superconductivity, nano-diamonds, oil sand extraction, and electronic kinetic sculpture. De Mayo and his wife, Judy, have two adult children and one grandchild.

'Buffalo Bill', plate steel, welded, 2019.

Chapter 1

Introduction to waves

We as humans receive the vast majority of our sensory perceptions through our eyes and ears. In this book, we will examine in a non-technical manner some of the everyday physics behind hearing and vision. This will help us understand more about ourselves and about our physical environment. The mathematics involved will be minimal and simple.

In our attempt at gaining an awareness of the everyday physics of sound and light, we will first examine waves and vibrations. We will see how we can detect sound and light waves and how we can obtain information from them. In later chapters, the parts of the eye and ear are detailed. After a discussion of wave properties, we will cover the perception of sound and light, including the subjects of music and color.

1.1 Vibrations and waves

Our vision and hearing work when our eyes and ears receive light and sound waves. Our brains convert these into information. It is important for us to know exactly what a wave is, and how our eyes and ears detect waves.

What is a wave? One of the simplest waves to visualize is the water wave formed by dropping a pebble in a still pool, figure 1.1. From this we can see one of the main properties of waves: they all have their origin in something moving. Here we see that the motion of the pebble into the water causes a disturbance which spreads out as time proceeds. This disturbance is the wave.

Another example of a wave is shown in figure 1.2. The hand giving a sudden up and down motion to the string causes a wave, which travels along the string. Both the water wave of figure 1.1 and the string wave of figure 1.2 are pulse waves. Compare these to sustained waves, as shown in figure 1.3. The hand continues its up and down motion over and over, forming the wave. If the hand moves up and down at a regular rate, that is, the same number of motions per second, then we will obtain

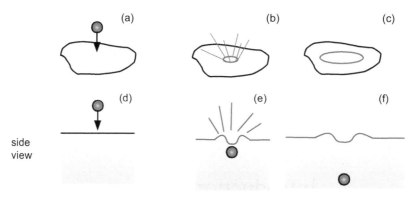

Figure 1.1. (a)–(c) A small stone is dropped into a still pond and a wave is formed. (d)–(f) The side view.

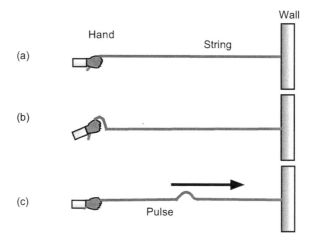

Figure 1.2. A string forms a *pulse wave* when the hand quickly moves up and then down.

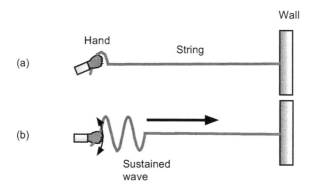

Figure 1.3. A continued motion of the hand forms a *sustained wave*.

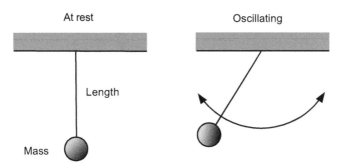

Figure 1.4. A simple pendulum illustrates periodic motion.

a *periodic wave*. If the hand moves up and down in an irregular manner, then it produces a *non-periodic wave*.

Periodic waves are perhaps the most interesting waves. They are common because there are many ways to produce periodic motion and the resulting waves. For example, the simple pendulum of figure 1.4 exhibits periodic motion.

We define the period as the time necessary for one complete vibration. The formula for the period, T is

$$T = 2\pi \sqrt{\frac{L}{g}},$$

where L is the length of the pendulum string and g is the acceleration due to gravity. Gravity is what pulls the pendulum downward; the stronger the gravity, the smaller the period and the faster the pendulum will swing.

Clockmakers have used this fact for centuries to construct accurate timing mechanisms and clocks. In the case of grandfather clocks, whose period is 2 s, the length of the pendulum is about 1 m. When you add the mechanism at the top and the space at the bottom for the weights, the clock is over six feet tall. This may be why there are fewer grandmother clocks—most grandmothers are well below six feet tall!

We can construct the analog of a simple type of wave as shown in figure 1.5.

If we take a look at the pattern made by the sand from above, it will look like figure 1.6.

We call this a *sine wave*; compare this to the sustained wave of the string in figure 1.3(b). We call any vibration which produces such a sine wave *simple harmonic motion*. Notice that as the pendulum makes one complete vibration, we obtain one 'piece' of the sine wave, as shown in figure 1.7.

One wavelength is the distance from one point on a wave to the next similar point. For example, one wavelength is the distance from one peak to the next peak, or from one trough to the next trough. The wavelength is related to the inverse of the period. Speeding up the vibrations means a shorter period and a shorter wavelength, figure 1.7(b).

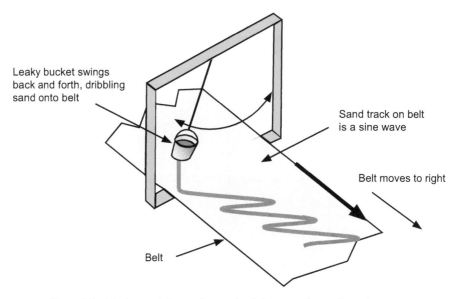

Figure 1.5. A leaky pendulum and a moving belt are used to make a sine wave.

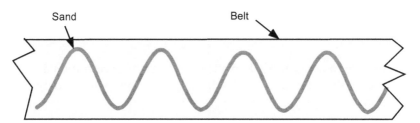

Figure 1.6. A top view of the belt shows the sand sine wave.

The mathematical relation between the wavelength and period is:

$$\frac{W}{T} = \text{a constant},$$

where W is the wavelength and T is the period. Another useful parameter of periodic waves is their *frequency*, that is, the number of vibrations occurring per unit of time:

$$f = \frac{1}{T}$$

where f is the frequency. If the period is 1 s, there is one vibration per second and the frequency is one cycle per second. If the period is 10 s, then the frequency is 1/10 of a cycle per second. Since 'cycles per second' is a little cumbersome to write and say, we use the term hertz, or Hz instead. This unit is named after Heinrich Hertz (1857–1894), a pioneer in the field of radio waves. As an example, consider a frequency of 100 cycles per second = 100 Hz; this frequency represents a period of 1/100 s.

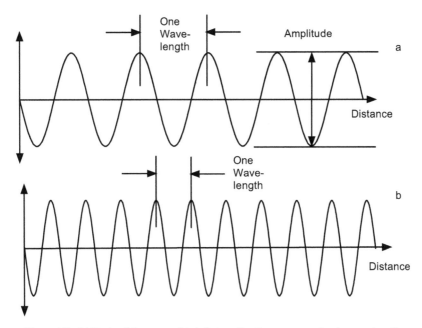

Figure 1.7. (a) Parts of the wave. (b) A faster vibration means a shorter wavelength.

All periodic waves do not have to be sine waves. For example, figure 1.8 shows some common periodic waves all with the same wavelength.

The last wave in figure 1.8 is called a *periodic complex wave*. There are many, many waves in this category, as we will see.

Some examples of non-periodic waves are shown in figure 1.9.

It is interesting to compare the side view of a sine wave as seen in a snapshot, where we see a whole segment of the wave, and the time development of one point on the wave, as we would observe by looking at one point on the string of figure 1.3(b), figure 1.10.

Each little piece of the string will have the same variation in amplitude as the wave moves down the string. Furthermore, the up and down motion of each point is that of simple harmonic motion.

Examples of different waves show some other features of waves. So far, we have considered water surface waves and a rope flipped by hand. The sand-dribbling pendulum was mainly introduced as an analogy, that is, a way to graphically generate a sine wave. The water surface wave can be a pulse, as in figure 1.1, or can be sustained as on the ocean, figure 1.11.

As the wave moves to the left, the bottle moves up and down but not sideways. The portion of water holding up the bottle moves up and down as the wave moves past it. It is the combined up and down motion of the various 'pieces of water' that allow the wave to propagate. In the flipped rope, the wave is propagated as sections of the rope move up and down. This brings up an important property of waves: waves transmit energy but not matter. More will be said about this later.

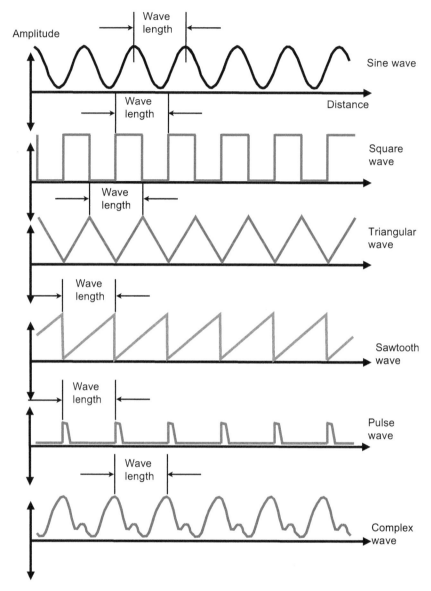

Figure 1.8. Various periodic waves.

In the water surface wave and in the flipped rope wave, the up and down motion of the parts of the water and rope allow the wave to be transmitted. Whenever the vibrations of the wave are perpendicular to the direction of propagation of the wave, as in these two examples, we have a *transverse wave*. One can compare and contrast this to the *longitudinal wave* shown in figure 1.12 for a spring.

Each loop of the spring moves at most only a small distance back and forth along the direction of propagation of the wave. The wave being transmitted is a series of regions where the spring is compressed and stretched. Let's take a snapshot of the

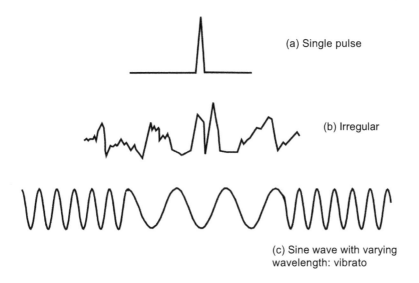

(a) Single pulse

(b) Irregular

(c) Sine wave with varying
wavelength: vibrato

Figure 1.9. Some non-periodic waves: (a) a single pulse, (b) an irregular wave and (c) a sine wave with varying wavelength (vibrato).

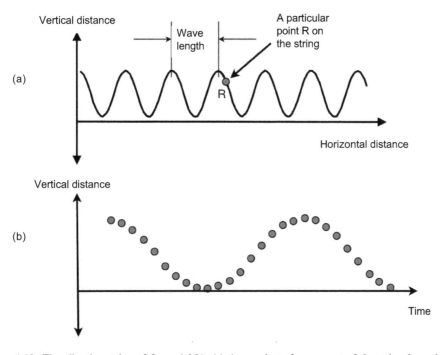

Figure 1.10. The vibrating string of figure 1.3(b). (a) A snapshot of a segment of the string from the side showing a particular point R. (b) The height of the point R as time progresses.

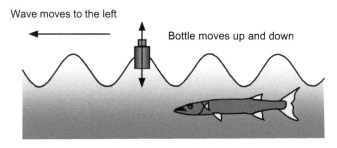

Figure 1.11. A bottle and a surface wave.

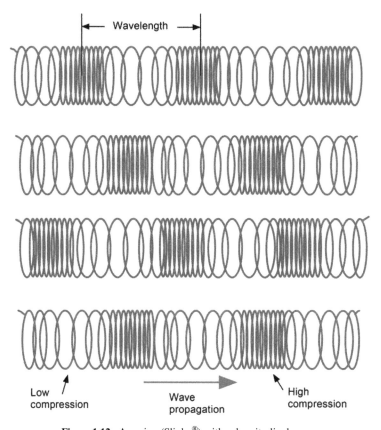

Figure 1.12. A spring (Slinky®) with a longitudinal wave.

spring as a wave is being transmitted along it and plot the number of spring loops per centimeter along the spring, figure 1.13.

Sound is a type of longitudinal wave. Instead of spring-loop variations, regions of high and low air pressure are transmitted, as shown in figure 1.14.

Each air molecule moves back and forth along the direction of the wave (to the left and right in the figure). The motion for each molecule is very small. This is all that is needed for the wave to propagate. These very small 'air compression waves' are what

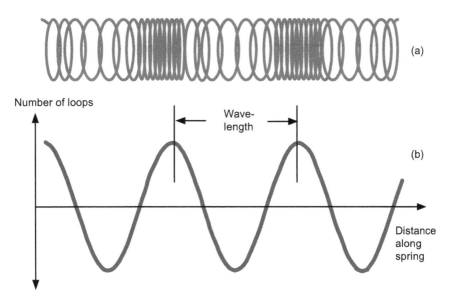

Figure 1.13. The spring and its loops. (a) The spring with a wave being transmitted. Notice the bunched-up loops. (b) The number of loops per unit of distance along the length of the spring versus the distance along the spring. Note that this is a sine wave.

our ears sense. Our ears can detect pressure levels of as little as 10^{-12} W m^{-2}, about the pressure generated by a gnat's wings. The arms of a tuning fork (called 'tines') vibrate with simple harmonic motion, generating a sine wave of air pressure, figure 1.14(b). This is why we hear a 'pure tone' from the tuning fork. The sound wave of a source playing the same note, say an oboe, would look like figure 1.14(c). The wavelength, period and frequency are the same, but now we have a complex periodic wave.

1.2 Energy and information

1.2.1 Energy

Waves transmit energy, not matter. Energy can be defined as the capacity to do work. Usually when something is moving, work can be extracted. For example, a running river can provide mechanical energy if a water wheel is placed in it, figure 1.15. In the case of our eyes and ears, we need the information sent by the sound or light wave. The variation of the waves' frequency and amplitude contains the information. The energy of the flowing water provides the work necessary to rotate the wheel. In a similar way, energy can be extracted from sound waves. Our eardrums harness the compression and rarefaction of sound waves as shown in figure 1.16. As we will see, the sound wave provides the energy to stimulate the sound receiving cells in our ears. But it is the variation in the intensity and frequency of the sound wave that we are interested in. We obtain the energy that is necessary to run our bodies from food in the form of chemical energy. Some other sources of energy are sunlight, wind, fossil fuels (gas, coal, oil), wood, hydropower, tidal flows

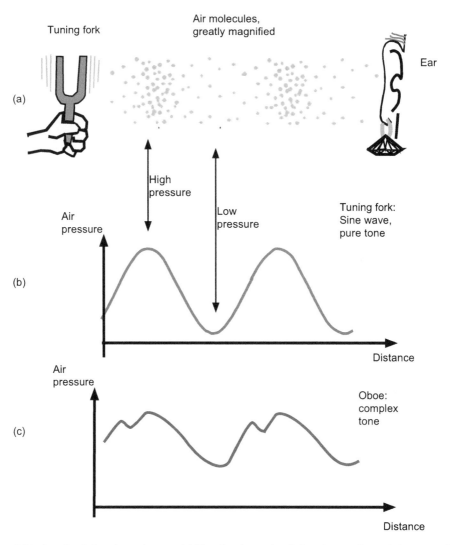

Figure 1.14. A tuning fork and sound waves. (a) The vibrating tuning fork and a sound wave being transmitted and received by an ear. (b) The density of the air molecules along the direction of the wave's transmission, versus the distance from the fork. Note that this is a sine wave. (c) A similar plot from a different source, say an oboe. Even though the wavelength is similar, the shape is not sinusoidal and it sounds different. Your ear is capable of distinguishing between the two different sound waves—and hundreds of thousands of other complex waves.

and ocean currents. All of these come from the Sun, directly or indirectly. Additionally, we have nuclear and geothermal energy resources.

1.2.2 Transducers

We can think of each of our five senses (touch, taste, hearing, seeing and smell—we can also sense heat and cold) as a different type of energy receptor. For example, the

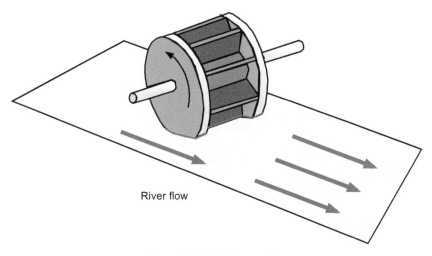

Figure 1.15. Work from a river.

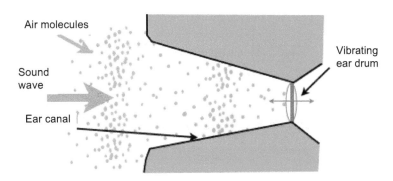

Figure 1.16. Energy from a sound wave.

sense of touch detects mechanical energy. It is somewhat more fruitful, however, to think in terms of transducers, rather than energy receptors. A transducer is a device which changes one form of energy into another form, as shown in figure 1.17(a). We see in figure 1.17(b) how the eye is really a transducer. Light energy strikes the receptor and nerve impulses travel out of it to the brain. The nerve impulses are electrochemical pulses. These tell the brain that the receptor has been activated.

Energy comes into the picture in two ways. First, the energy in the light wave sets off a chemical trigger in the receptor. This causes a much larger release of energy, as shown in figure 1.18(a). One example of a trigger device is the pistol, figure 1.18(b) and (c). When we squeeze the trigger far enough, the hammer flies forward, causing the firing pin to strike the primer. The primer is extremely sensitive to such an impact. A small explosion results which sets off the stored chemical energy of the gunpowder. The resulting hot gases force the bullet out of the barrel of the pistol with great speed—and energy. The relatively small amount of input energy (the trigger squeeze) results in the release of a large amount of energy of an entirely

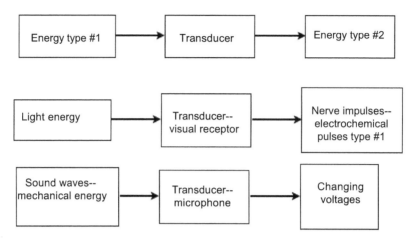

Figure 1.17. A transducer changes one form of energy into another (a). The eye, for example, changes light energy into electrochemical pulses (nerve impulses) (b). A microphone changes the mechanical energy of a sound wave into a changing voltage (c).

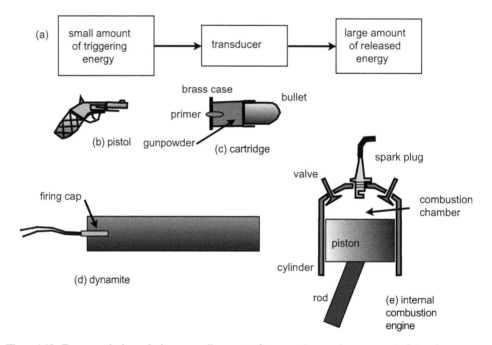

Figure 1.18. Energy and trigger devices: a small amount of energy releases a large amount of stored up energy. (a) A schematic. Examples: (b) pistol, (c) gun cartridge, (d) dynamite and (e) internal combustion engine.

different type (the moving bullet). The pistol is a type of transducer. We release a much larger amount of energy of one type than the type we put in. We can do this because the pistol contains the stored energy (gunpowder) of the cartridge. Figure 1.18(d) and (e) show two other examples of trigger devices.

The eye, the pistol and the microphone are all transducers. In our sense organs, the receptors act as trigger devices: a small energy input results in a large energy output. The receptor cells of the eye and ear use the trigger stimulus (light and sound) to release stored chemical energy. Energy comes into the picture in a second way. Our bodies obtain chemical or food energy by eating and drinking. Part of this goes to setting up and maintaining the energy storage systems of the receptors. This is a complicated process, with many factors involved. Basically, however, our senses are transducer-type trigger devices.

1.2.3 Information

Our senses thus do not collect energy for their own sake. They use the various forms of energy as inputs to inform our brain about the environment that our body finds itself in. Is it too hot or too cold? Is something mashing our foot? Is what we are eating spoiled? Is there a skunk nearby? These are things that our sense organs along with our brain can tell us. The senses really gather information—knowledge about our surroundings. Energy and changes in energy convey this vital information to us. In this book we are mainly concerned with sound and light, and our senses of hearing and vision. Through these we obtain the most information and through them we convey information to others.

Chapter 2

The ear

2.1 Parts of the ear

The human ear is a truly remarkable organ. It allows us to detect sounds as soft as those made by gnat's wings and as loud as a cannon's roar. Additionally, we can pick a familiar voice out of the hubbub of a crowded room and we can discern the bassoon in a full orchestra. The normal human can hear frequencies as low as 20 Hz and as high as 20 000 Hz. We can distinguish 1500 individual frequencies and 400 000 separate combinations of sounds. Even while we sleep, we hear. A baby's cry can rouse the mother even though there are louder background noises occurring while she is sleeping. In this chapter, we will examine the main parts of the ear and the fundamentals of its workings. We will leave until later the subjects of auditory perception, musical sounds, tone and pitch, auditory illusions and the musical scale.

Figure 2.1 shows the main parts of the ear.

We can divide the ear into three parts: the outer ear, the middle ear and the inner ear. The outer ear is the visible part; we attach earrings to its earlobe. The pinna collects sound and focuses the waves into the auditory canal (meatus). The meatus ends at the eardrum. The eardrum vibrations are tiny. While receiving normal conversational sounds, the eardrum moves only 1/100 000 000 (10^{-11}) cm. The middle ear contains the hammer (malleus), anvil (incus) and stirrup (stapes). These three small bones, the 'ossicles', transmit the vibrations from the eardrum to the oval window, figure 2.2.

These bones change the weak air vibrations over a wide area to strong vibrations over the smaller area of the oval window, as shown in figure 2.3(a).

Also in the middle ear is the Eustachian tube, which equalizes the pressure between the middle ear and the outside air. The inner ear contains the three mutually perpendicular semicircular canals. These play no part in hearing, but instead allow us to maintain the balance and equilibrium necessary for sitting, standing, walking and running. The oval window is the connection between the stirrup and the cochlea, also located in the inner ear. The cochlea is coiled like a snail and is only about as big

doi:10.1088/978-0-7503-3207-1ch2

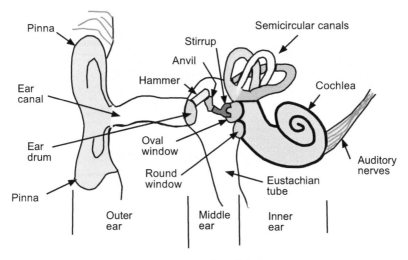

Figure 2.1. A cross-section of the human ear.

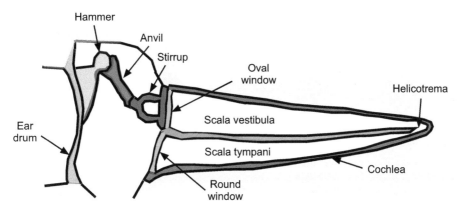

Figure 2.2. The middle ear and the cochlea unrolled (schematic).

as your little finger's tip. This marvel of miniaturization is well protected by being buried deep in the petrous portion of the temporal bone, one of the hardest in the entire body.

The sound vibration goes from the stirrup through the oval window and into the scala vestibuli, figure 2.4. Figure 2.5 shows a cross-section of the cochlea.

As the wave travels along the cochlea, it excites sections of the basilar membrane. High frequency sounds excite regions near the base of the cone. Low frequency sounds excite the region near the helicotrema, or apex of the cone. The sound vibrations pass out of the cochlea through the scala tympani and then through the round window into the middle ear. The location of the excitation on the basilar membrane tells the brain what the sound's frequency was. The degree of excitation reveals the intensity of the original sound.

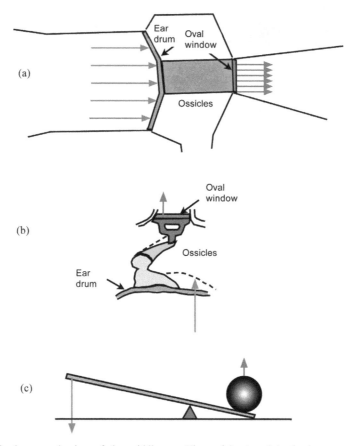

Figure 2.3. The lever mechanism of the middle ear. The ossicles translate the large area, low pressure movement (pink) of the eardrum into smaller but more powerful vibrations (blue) at the oval window: (a) schematic; (b) a diagram. A similar thing happens with a lever (c): a small force (pink) acts over a large distance to produce a large force over a small distance (blue).

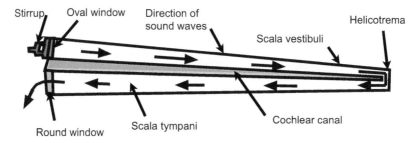

Figure 2.4. A diagram of an unrolled cochlea showing the pathway of the sound: the stirrup transmits the sound vibrations into the scala vestibuli through the oval window, around the helicotrema, back down the scala tympani and out through the round window into the Eustachian tube.

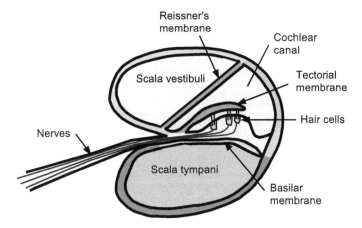

Figure 2.5. A cross-section of the cochlea.

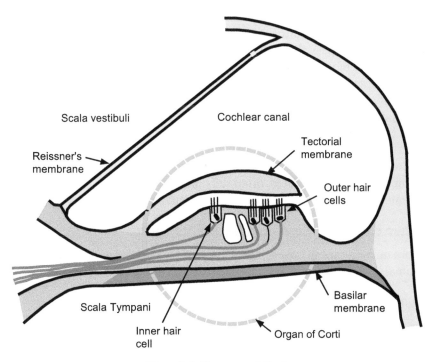

Figure 2.6. The organ of Corti.

Named for Italian anatomist Marquis Alfonso Giacomo Gaspare Corti, 1822–1876, the organ of Corti, shown in figure 2.6, transforms the sound from mechanical vibrations into nerve impulses. Georg von Békésy, 1899–1972, advanced our understanding of this mechanism, for which he won a Nobel Prize in 1961. As the basilar membrane vibrates, the tectorial membrane slides sideways across the hair cells, causing them to send a message along the nerves to the brain. In a similar way,

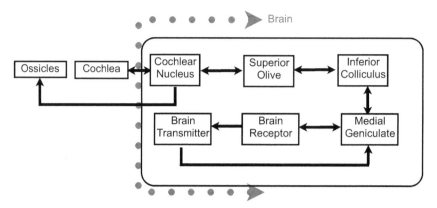

Figure 2.7. The auditory pathways to the brain (schematic).

you can use a pencil tip to excite the hairs along the back of your arm. For each ear, there are approximately 3500 inner hair cells, 20 000 outer hair cells and 30 000 nerve fibers forming the auditory nerve bundle.

The nerve signals pass through several relay stations on the way to the brain's auditory complex, figure 2.7. These stations also serve to relay messages back to the ear. For example, when an especially loud sound hits the ear, the brain sends messages to the eardrum, which then changes its shape to increase its stiffness. Also, the ossicles adjust themselves into a less sensitive arrangement. This adjustment cannot take place instantaneously, however. Sudden, intense sounds such as gunfire can permanently damage the ear.

Once the sound messages reach the brain, exceedingly complex interactions occur. The hearing center is connected intimately with areas of the brain which control sound memory, language formation, speech, visual memory and emotions.

2.2 Other ears and the non-acoustic labyrinth

We can obtain a picture of how our ears evolved by examining the hearing devices of other animals.

2.2.1 Statocysts and equilibrium

In certain very rudimentary animals such as the medusa phase of the *Obelia* jellyfish, a simple organ is used to maintain balance (figure 2.8).

The statolith is a small stone-like object which rests on the bristles of the sensory cells ('hair cells'), which in turn are located in the bottom of the statocyst chamber. When the jellyfish tilts, the statolith moves and excites the sensory cells by moving the bristles.

2.2.2 The non-acoustic labyrinth

Our sophisticated organs of hearing developed from these simple balance detectors. In addition to hearing, humans also must maintain balance for every activity except reclining. The *non-acoustic labyrinth* tells us our orientation (up–down) and also tells

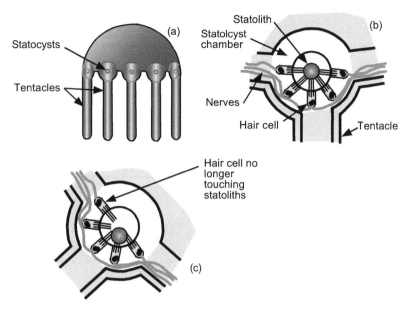

Figure 2.8. The medusa phase of the *Obelia* jellyfish: the statocysts may be the ancestors of the human ear. (a) The *Obelia* jellyfish, (b) a cross-section and (c) the effect of tilt: the statolith moves and excites the hair cells.

us whether we are accelerating (changing our speed). It consists of the three semicircular canals next to the cochlea, figure 2.1. Each canal has a small bulbous ampulla. Two small attached chambers, the utriculus and the sacculus (figure 2.9) complete the labyrinth.

When the head is tilted, the otoliths embedded in the gelatinous membrane in the region of the sacculus/utriculus (called the macula) (figure 2.9(b)) shift the membrane, exciting the hairs in the hair cells, which then send signals to the brain telling it the head's orientation.

When the head *accelerates* to the left (green arrow) (figure 2.9(c)), the endolymphatic fluid shifts to the right (red arrow), distorting the gelatinous cupula and exciting the hairs in the hair cells, which send a message to the brain via the nerves. In the glass of water on the right, the *acceleration* of the glass and water shifts the water to the right (red arrow). When the *acceleration* stops, the water and endolymphatic fluid shift back to the right, even though the glass/head is still moving at a constant velocity. The non-acoustic labyrinth detects the head's orientation and *acceleration*, not the velocity.

The eyes also assist in maintaining equilibrium, as do small sensory endings in the muscles and tendons called proprioceptors. These are sensitive to changes in tension. An experiment to demonstrate the importance of vision to balance, try the following. First, stand up; then lift one foot off the floor. Balancing is easy. Second, close your eyes before lifting your foot. Without vision, balance is very difficult. Next, before lifting your foot and before closing your eyes, focus your vision on a particular object in your line of sight. *Now* close your eyes and lift your

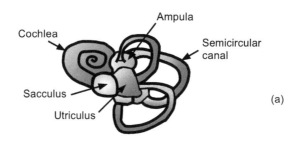

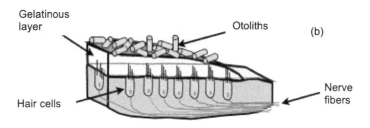

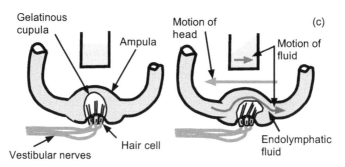

Figure 2.9. The non-acoustic labyrinth. (a) An overall view with the cochlea. (b) A cross-section of the sacculus/utriculus in the region called the macula. (c) A cross-section of one of the semicircular canals in the region of the ampula. Also shown is a glass of water, stationary and moving to the left.

foot. Your balance should be a lot better, although not as good as with your eyes open! The reason that there are *three* semicircular canals is that this is the minimum number needed to identify motion in all directions. However, the planes of the semicircular canals must be mutually perpendicular, as shown in figure 2.10.

This arrangement ensures that for any direction of motion, at least one of the canals in each ear will function. When movement occurs, fluid flows in the canal opposite to the direction of motion. If the movement is more or less horizontal, everything proceeds in the manner to which we are accustomed. However, excessive vertical or swaying motion can cause very unpleasant effects: sea sickness, car sickness or air sickness. No-one knows the exact connection between the non-acoustic labyrinth and the stomach, but one surely exists (at least indirectly)!

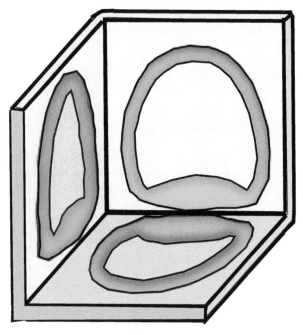

Figure 2.10. The mutually perpendicular arrangement of the semicircular canals. The plane of each is perpendicular to the planes of each of the other two.

2.2.3 The lateral line

Another step on the evolutionary ladder of hearing development is perhaps the lateral line of all fish and many amphibians. The lateral line of the completely blind cave fish allows it to move about in total darkness without hitting obstacles. Fish are able to stay in schools by detecting the pressure waves set up by the other fish in the school. The African clawed frog has a lateral line that is extremely sensitive to movements of the very muddy water in which it lives, figure 2.11.

As shown in figure 2.11, the lateral line contains gelatin-like cupulae which contain hair cells. These cells are excited by movements of the cupulae, which in turn are caused by minute pressure changes in the water. Note the similarity to the human inner ear (organs of Corti) and the lateral line (sensory hair cells).

2.2.4 Fish air bladders

The lateral line allows the detection of water movement. Some fish can also detect water-borne sound waves: vibrations in the water. The vibrations pass from the water, through the flesh of the fish into its air bladder. There the vibrations are changed into air vibrations. The bladder is a gas-filled cavity which allows the fish to maintain its desired depth in the water, sort of like a balloon. It has connected to it four small bones (Weberian ossicles); these and the bladder make up the fish's middle ear. In the manner of our ears, these bones transmit the vibrations to the

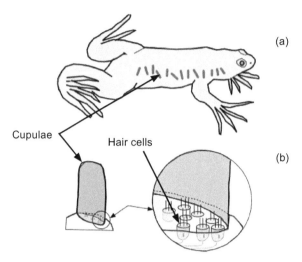

Figure 2.11. (a) The African clawed frog, showing the lateral line. (b) The cupulae of the lateral line are sensitive to minute pressure differences.

inner ear, which contains a labyrinth (figure 2.12). The function and construction of the fish and human middle ear and inner ear are remarkably similar.

2.2.5 Amphibian ears

The ear of frogs and other amphibians offer an interesting insight into how nature works. Amphibians spend the early part of their lives as water creatures, complete with gills. Then they develop lungs and emerge from the water. For example, the frog egg hatches into a tadpole, which lives entirely in water, figure 2.13(a). Later it develops into a land-living frog. As a part of this transformation, the hearing organs also change considerably. The tadpole uses the developing lungs as resonators to pick up underwater vibrations (in a manner similar to the use of the air sac by fish). The frog, on the other hand, has a large eardrum to detect sound waves in the air, figure 2.13(b). A major change occurs in the transformation from tadpole to frog. This change allows the frog to hear the air vibrations of his new environment, while the tadpole could hear only the vibrations in water. Note that there is no cochlea.

The eardrum as developed by the amphibians was a major evolutionary step up in hearing. As animals emerged from the sea and traveled about on land, their ears allowed them to detect the plethora of vibrations present in the air. The land animals quickly attained the ability to generate their own sounds, for various purposes, as we can tell from the cacophony of sounds that pour forth from the frog inhabitants of any bog on a summer's evening.

2.2.6 Bird ears

Frogs hear very well, as we can tell from the complexity of frog songs. (It is doubtful that they would sing what they couldn't hear.) However, birds hear much better, due

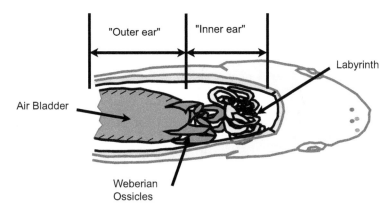

Figure 2.12. The fish hearing system.

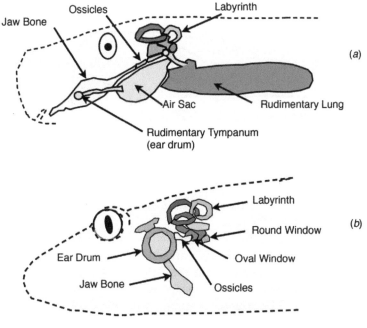

Figure 2.13. (a) Tadpole. (b) Frog.

to the important addition of the cochlea, figure 2.14. The beauty of a mockingbird's concert indicates just how well the ears of birds must be able to operate. Birds lack external ears due to the necessity of streamlining during flight, figure 2.15.

2.2.7 Insect ears

The ears of insects are very simple but very effective. They are located on different parts of the insect's body, depending on the species. Mosquitoes hear through their

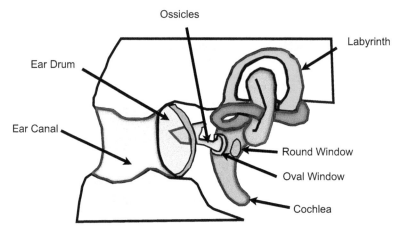

Figure 2.14. The bird ear. Note that the cochlea is only slightly bent, not spiraled as in humans.

Figure 2.15. A buzzard head. The lack of head feathers allows us to observe the ear hole.

antennae; moths and butterflies may have their ears at the bases of their wings. The katydid has ear-slits on the lower part of each of its front legs. Many insects, such as the grasshopper, have ears on their thorax. As shown in figure 2.16, for insects, the eardrum forms an outer surface of an air-filled cavity and has nerves attached directly to it. The counter-tympanum allows the sound waves to escape from the cavity.

2.2.8 Other mammal ears

It would be interesting to investigate in detail the various mammal ears. Since that would take more space than is available here, let us just mention a few of the outstanding ones. Consider the lowly bat, which navigates in the dark by emitting high frequency chirps and accurately receiving the echoes as they are bounced off objects into which the bat might collide. Also note the dolphin and other water-dwelling mammals, which use a similar system of sonic pulses in water

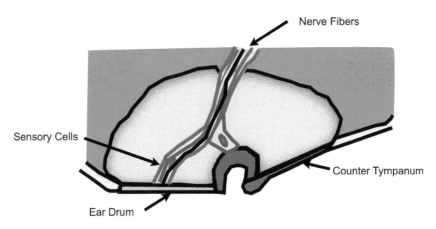

Figure 2.16. A moth ear, top view.

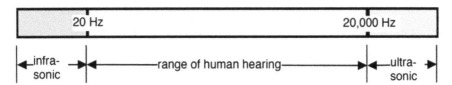

Figure 2.17. The sound spectrum.

(sonar). Small nocturnal mammals, such as the rabbit, couple large exterior ears to sensitive inner parts to achieve the remarkable levels of hearing necessary for the detection of approaching predators. Many mammals have wider frequency ranges of hearing than humans, as well as much more sensitive hearing.

2.3 The sound spectrum

To accurately compare these frequency ranges, we need to find out a little about the sound spectrum, figure 2.17. The very low frequencies, *infrasonics*, are not detected directly by humans, although recent findings show that infrasonic sound waves can cause disorientation and collapse in humans. One way that we can be exposed to these waves is by the buffeting that occurs when we drive with one window open in a car. It is feared that serious accidents result from the effects of such infrasonic vibrations.

Since much of this book deals with the audible range, 20–20 000 Hz, for the time being we will skip over this to the *ultrasonic* range. As shown in figure 2.18, many animals can produce and hear ultrasonic waves that are imperceptible to humans. Ultrasonic waves can be used in cleaning small parts such as mechanical wristwatch mechanisms, in testing metal parts by detecting the echoes from cracks and other imperfections, and even in making remarkable 'pictures' of unborn babies as in figure 2.19.

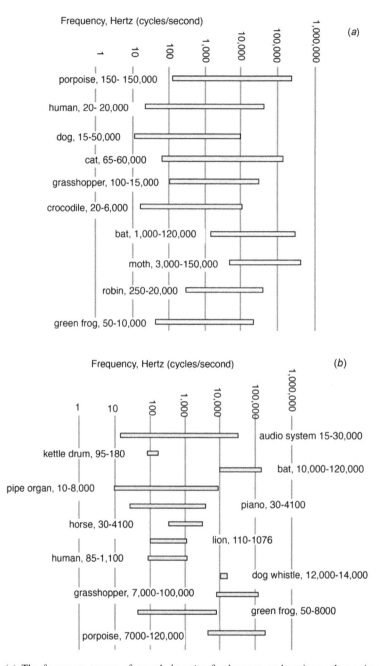

Figure 2.18. (a) The frequency ranges of sound *detection* for humans and various other animals. (b) The frequency ranges of sound *production* for humans and various other animals and instruments.

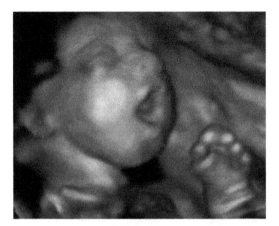

Figure 2.19. An ultrasound image of an unborn baby.

We need to be aware of the fact that sound waves exist in the frequency ranges that are undetectable by us, except with electronic devices.

2.4 Aids to hearing

Hearing is very important to us and losses in hearing have many causes. It is not happenstance that we have two ears to ensure that if one is put out of commission, we do not become totally deaf. Along the same lines, our middle and inner ears and our non-acoustic labyrinths are surrounded by the densest bones in our bodies.

As we age, our hearing gradually diminishes (presbycusis) and any number of hearing maladies crop up. Is it then any wonder that a multitude of hearing aids and devices have been and are being developed? In the following some examples are discussed.

2.4.1 Ear trumpets

Ludwig van Beethoven was going deaf in the 1810s and probably used an ear trumpet to boost his hearing, figure 2.20. The device concentrates incoming sounds in a manner as in the human ear canal, but on a larger scale. Just as in the case of binoculars, the aperture of the trumpet is much larger than for the ear, and the sound is intensified by the time it hits the eardrum.

2.4.2 Digital hearing aids

Hearing aids have come a long way since Beethoven's ear trumpet, having taken advantage of the same technology that's given us the smart phone and such. Digital and analog hearing aid versions are compared in figure 2.21. Current hearing aids compensate not only for hearing loss, but they are also capable of echo cancellation, reduction of background noise and enhanced discernment of voices. Therapeutic functions such as treating tinnitus, a serious ringing of one's ears, are also available. Not only are capabilities enhanced, but size and cost are reduced.

Figure 2.20. Drawing of Beethoven's probable ear trumpet, used to boost his failing hearing ability. From an undated print based on a painting by Carl Jaeger (1833–1887).

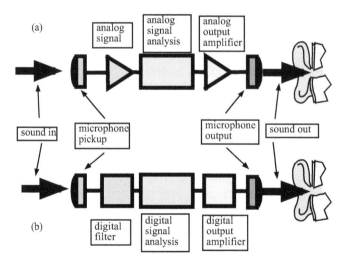

Figure 2.21. Diagram of analog (a) and digital (b) hearing aids.

Digital hearing aids amplify and digitize the input audio signal in the A/D (Analog to Digital) converter. The real work goes on in the next part of the process: the signal analyzer. Advanced software can filter and adjust the digitized signal, then send it on to the digital to analog, D/A, converter. Finally the audio signal goes to the eardrum.

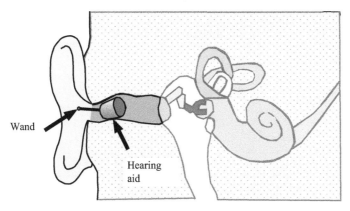

Figure 2.22. The hearing aid is entirely in the ear canal. The wand assists in extraction and re-insertion of the hearing aid.

Some models of new digital hearing aids can be wirelessly programed and controlled by tablets or smart phones. Television sets and other electronic devices can also be communicated with via Blue Tooth or Wi-Fi. If you have one hearing aid for each ear, the two can work together with signals right through your head—a real improvement. Also available are implantable rechargeable batteries.

Some hearing aids are small enough to fit entirely inside the ear canal, figure 2.22. First, a wax impression of the ear canal is 3D scanned. Software generates a virtual 3D model whose outer dimensions are just slightly smaller than those of the ear canal. Technicians arrange the electronic components of the hearing aid to fit inside the virtual model. The last step is to 3D-print the virtual model and insert the electronic components. These components include the pick-up microphone, the digital electronic chip, the battery, and the output microphone. For cleaning and battery replacement, a small wand helps in removing and replacing the hearing aid from and into the canal.

2.4.3 Cochlear implants

Cochlear implants stimulate the cochlea directly, figures 2.23 and 2.24. The external ear hook contains sophisticated electronics along with a connection to the externally placed transmitter.

Magnets hold the transmitter and internal receiver in place while electronic signals are exchanged through the skin. These signals go through the stimulator cable straight into the cochlea's scala tympani. The cable lies directly under the hair cells, basilar membrane, and auditory nerve fibers, figure 2.24. The stimulator cable stretches out all along the cochlea; thus, the cable touches all regions of the tympanic membrane and all parts of the sound spectrum are transmitted. In some versions of the device, the batteries of the internal receiver can recharge

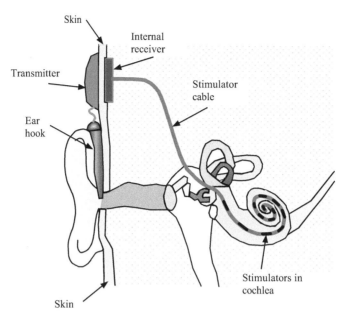

Figure 2.23. Direct stimulation of the cochlea using an implanted device.

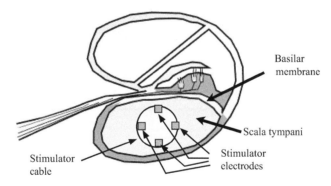

Figure 2.24. Cross-section of the cochlea showing the implanted stimulator cable and electrodes. Figure 2.5 labels all of the parts.

right through the skin. Figure 2.25 shows a diagrammatic side view of the direct stimulation system.

This system is highly effective for certain types of profound hearing loss.

2.4.4 Gene therapy for hearing loss

Some genetic defects result in damaged hair cells and a concomitant loss of hearing. Gene therapy can treat the condition. Corrective DNA is attached to viruses such

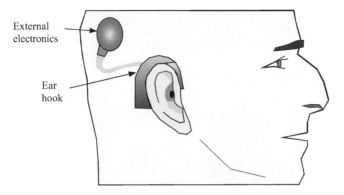

Figure 2.25. Side view of direct stimulation system.

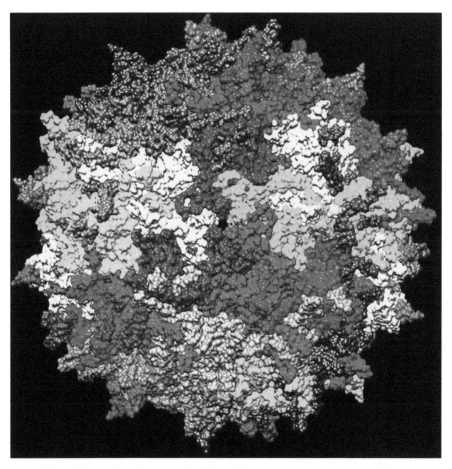

Figure 2.26. The AAV virus is used to transmit DNA which restores defective hair cells. This image has been obtained by the author from the Wikimedia website where it was made available by Jazzlw under a CC BY-SA 4.0 licence. It is included within this article on that basis. It is attributed to Jazzlw.

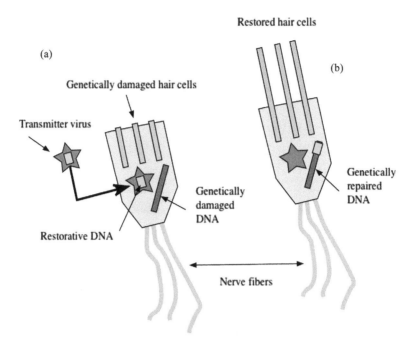

Figure 2.27. Simplified version of how to repair hair cells. (a) Damaged hair cells with defective DNA and the transmitter virus. (b) The transmitter virus inserts restorative DNA, genetically repairing the hair cells.

as AAV8, figure 2.26. The virus is injected directly into the cochlea; and the defective hair cells are replaced, figure 2.27. Preliminary results at Columbia University are promising.

We humans have two main methods of communicating with each other: auditory and visual. Recent progress in addressing hearing problems is very encouraging. As we shall see later, there are similar advancements with visual problems.

IOP Publishing

The Everyday Physics of Hearing and Vision (Second Edition)

Benjamin de Mayo

Chapter 3

The eye

Your eyes are your window into the world. You can gain a sense of the importance of your vision easily by closing your eyes for half an hour (no napping, please). If you gain nothing else by reading this book than a heightened appreciation of your eyes, then your experience will have been worthwhile.

In this chapter, we will first take a look at each section of the electromagnetic spectrum and how each part affects us. Visible light, the segment that affects us the most, is but one small piece of the whole spectrum, which also includes gamma rays, x-rays, ultraviolet (UV) rays, infrared (IR) rays, microwaves (MWs) and radio waves. Then we will examine how the eye works, its parts and some interesting eyes of other creatures.

3.1 The electromagnetic spectrum

Just as the sound spectrum contains parts that are inaudible to us, there are parts of the light spectrum that are invisible. Light is one form of electromagnetic wave. Electromagnetic waves are transverse vibrations of electric and magnetic fields. They all travel through space at the same speed of 3×10^8 m s^{-1} (186 000 miles/s), figure 3.1.

Each type of electromagnetic wave has a different range of wavelengths and frequencies. The smaller the wavelength, the larger the frequency, and vice versa. All of the types together, from the shortest wavelength to the longest, make up the electromagnetic spectrum. Visible light occupies a very narrow range of the entire electromagnetic spectrum, figure 3.2. There is no hard and fast separation between the different parts of the spectrum and some overlap exists. Starting with the shortest wavelength, let us now look briefly at the parts of the spectrum. For each part of the spectrum we will note its wavelength and frequency range plus how it relates to our everyday life. It is helpful to use the concept of the photon, a quantum of electromagnetic energy. In analogy, we can think of a stream of water as a bunch of individual droplets and an electromagnetic wave as a stream of photons. Einstein

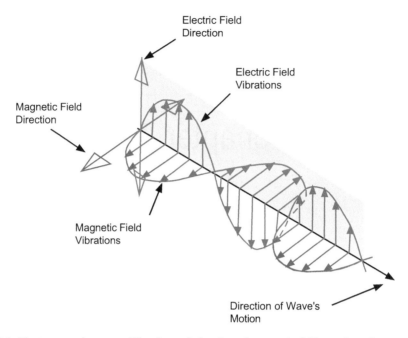

Figure 3.1. Electromagnetic waves. Vibrations of electric and magnetic fields produce these waves. The vibrations are perpendicular to the direction of the wave's motion (transverse).

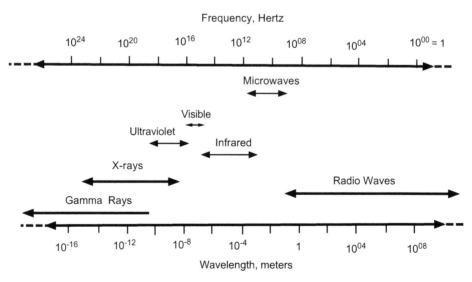

Figure 3.2. The electromagnetic spectrum. Notice that the scales are not linear.

first postulated the existence of photons to explain the photoelectric effect. When we shine light of certain wavelengths onto metal, electrons are emitted from the metal. Einstein theorized that the energy of each photon of light is directly proportional to its frequency f. In equation form, the energy of a photon, E, is $E = h \times f$, h is a number (called Planck's constant $= 6.63 \times 10^{-34}$ J s^{-1} in metric units) and f is the frequency of the photon in hertz (Hz). A photon striking the metal will knock out an electron if its energy is high enough, that is if its frequency is high enough. For this work, Einstein received the Nobel Prize in 1921.

3.1.1 Gamma rays

Gamma rays have the shortest wavelengths and thus the highest frequencies of all the electromagnetic waves. Therefore gamma ray photons are the most powerful of the electromagnetic radiations. Processes of nuclear decay produce gamma rays. Like fire, gamma rays can be very beneficial or very harmful. When gamma rays strike a living organism, damage can occur at the cellular level. Cells which are dividing are particularly vulnerable. Cobalt-60, a radioactive isotope of cobalt, emits an especially powerful gamma ray photon when it decays. 'Cobalt radiation treatment' is often used to destroy cancerous tumors. A cancer is a mass of tissue whose cells are multiplying out of control. In cobalt or other radiation treatments, the cancer cells are (ideally) killed in greater abundance than the surrounding normal cells. Gamma rays can, however, also cause cancer and as a result radioactive materials are very carefully controlled by governments. Radioactive materials have many uses besides cancer treatment.

3.1.2 X-rays

X-rays make up the next part of the electromagnetic spectrum; they have a slightly lower frequency and a slightly longer wavelength. The origin of x-rays and gamma rays is different. Gamma rays, as mentioned before, originate when radioactive nuclei decay. X-rays, on the other hand, are produced in atomic (not nuclear) transitions. Physically, the resulting photons are identical, except for their energies. Practically speaking, the difference in the method of generation of the photons is significant. For example, it would be difficult to use gamma rays to make dental x-rays, simply because of the relative difficulty of handling and arranging a radioactive source to produce the desired collimated beam. It would be even more difficult to make an x-ray generator that could be injected into the bloodstream to act as a radioactive (gamma ray) tracer.

As with almost everything else, x-rays can be put to good use, or they can cause great harm if misused. X-rays have the interesting (and useful) property of passing through some materials more readily than others. Since they will also activate photographic film or electronic detectors, medical and dental 'x-rays' are easy to make, figure 3.3. The x-ray beam passes through the flesh of the hand and into the film, activating the film. However, the x-rays are stopped by bone so that the film under the bones is not exposed. When the film is developed, the bones, nails, and even some cartilage are shown clearly. If an x-ray absorbing material is injected into some of the blood vessels, the blood vessels will also show up clearly. The dark side of x-rays is due to the fact that they, like gamma rays, can cause cancer. It is of paramount concern that our exposure to x-rays be minimized and much effort by health authorities is devoted to ensuring that no more than the required minimum exposure be received

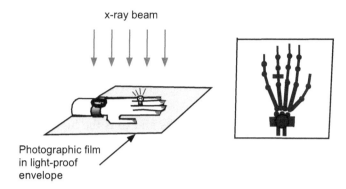

Figure 3.3. How x-rays are made. Note the ring and watch images in the x-ray.

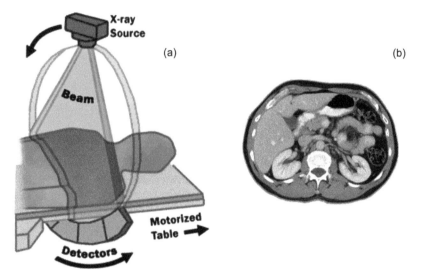

Figure 3.4. How a CT scan works. (a) A diagram of the process. (b) An example of a scan. https://www.fda.gov/radiation-emitting-products/medical-X-ray-imaging/computed-tomography-ct. Credit: U.S. Food and Drug Administration.

by the public. This has led to the discontinuance of once-standard practices, such as shoe-store x-ray machines for testing proper fit, and the lung x-ray surveys in the 1950s of high school students that attempted to detect tuberculosis cases.

Computed tomography (CT) scans detect multiple x-ray images electronically and store them in a computer, which then reconstructs them into a highly detailed image, figure 3.4.

3.1.3 Ultraviolet

UV rays, the next part of the spectrum, cause reactions when they hit our skin. One of these reactions is tanning, a darkening of the skin. The UV rays activate a pigment in the skin called melanin. If the exposure is too sudden and/or intense, the

UV rays cause sunburn. Suntan lotions attempt to block out the more energetic, UVA 'burning' rays (wavelengths of 2.9–3.2 × 10^{-7} m) while permitting the desirable UVB 'tanning' rays (up to 4 × 10^{-7} m). Vitamin D is also produced when UV rays strike the skin. However, too much exposure to UV rays can cause skin to become leathery and cracked; in extreme cases, skin cancer results.

Please be aware of the ABCDEs warning signs of melanoma, the worst form of skin cancer. A is for an Asymmetric border of a mole: if you draw a line through the middle of the mole, the two halves will not match. B is for Border: an early melanoma skin cancer tends to have uneven borders. C is for Color, which can be a number of shades of brown, black and red in melanomas. D is for Diameter, which is larger than a pencil eraser for melanomas. And finally, E is for Evolving: any change, in color, size, shape or elevation may indicate melanoma and a quick trip to consult a dermatologist is in order.

UV rays reach us from the Sun; our ozone layer blocks out much of this radiation. Partial destruction of the ozone layer, especially by chlorinated fluorocarbons, has been of great concern. An increase in the incidence of skin cancer will result if higher levels of UV radiation reach the Earth's surface; a more serious consequence to the vast numbers of the world's under-nourished would probably come about from the damage to agricultural plants that would also occur. The ability of UV, or 'black light', to make certain materials glow or fluoresce is used for special visual effects and for medical research.

3.1.4 Visible light

Visible light is the most interesting part of the spectrum for most of us, because this is the part that our eyes perceive. Most of the light that we use comes from the Sun. Figure 3.5 gives the output of the Sun for the parts of the spectrum. From this we

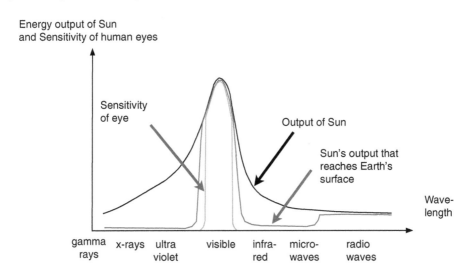

Figure 3.5. The spectral output of the Sun and the sensitivity of the human eye at different wavelengths. The portion of the Sun's output that reaches the Earth's surface is also shown.

can see that Mother Nature has made our eyes sensitive to that portion of the Sun's spectrum which is the most profuse. Note that visible light and radio waves are the only parts of the electromagnetic spectrum that reach the ground in any appreciable amount.

The visible light region of the electromagnetic spectrum occupies a very narrow part of the entire spectrum. When our eyes are exposed to the entire visible spectrum region evenly, we see white light. If we break up the white light with a prism, as shown in figure 3.6, we see that there is a smooth flow of colors, from red to violet. By common consent, the colors have the following wavelength regions:

Color	Wavelength region ($\times 10^{-7}$ m)
Violet	3.5–4.0
Indigo	4.0–4.5
Blue	4.5–5.0
Green	5.0–5.5
Yellow	5.5–6.2
Orange	6.2–6.7
Red	6.7–7.5

Notice that the colors' arrangement by wavelength can be remembered with the mnemonic Roy G Biv.

Some of the physiological effects of visible light are interesting. (1) The most important effect is probably photosynthesis: the use of light energy in the presence of chlorophyll to change H_2O and CO_2 into sugar. This is the basis for practically all life on Earth. (2) At the cellular level, visible light causes changes in such things as protoplasmic viscosity, permeability and the colloidal behavior of proteins. (3) Visible light rejuvenates UV-damaged microorganisms. (4) Phototropism allows plants to grow toward the source of light. (5) Visible light falling on the retina stimulates the pituitary-hypothalamus glands near the brain. Among other things this induces 24 h rhythms in animals and controls skin color in fish and amphibians.

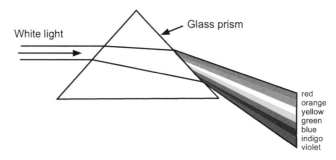

Figure 3.6. The prism separates white light into its colors.

3.1.5 Infrared

IR has wavelengths slightly longer those of visible light. Although we cannot see IR, we can feel it as heat striking our skin: almost 100% is absorbed. Supermarket managers use IR-generating light bulbs to keep pre-cooked foods warm. In bathroom ceiling fixtures, IR bulbs counteract the evaporative cooling of a wet bather as he towels off.

As shown in figure 3.5, much of the Sun's IR is absorbed in the Earth's atmosphere. An IR picture of the Sun taken from Earth's surface shows a dark disk. Above the Earth's atmosphere, the Sun would look bright in IR. Diseased vegetation reflects IR differently than healthy plants do. IR cameras can be used to distinguish between healthy and sick plants using drones, airplanes and Earth resource satellites.

Methane, CO_2, and H_2O vapor are labeled the greenhouse gasses, because they act like the glass of a greenhouse. In the case of an actual greenhouse, sunlight passes through the glass and is absorbed by the contents of the greenhouse. The contents then heat up and re-emit IR radiation. This re-emitted IR radiation cannot pass back through the glass and is therefore trapped in the greenhouse, which heats up.

This process is a nice way to heat up a greenhouse without using any fuel. A similar process occurs for Earth. Sunlight heats the Earth's surface and IR is radiated back toward space. However, the greenhouse gasses absorb the IR radiated from the Earth's heated surface. As in the original greenhouse, the heat is trapped and the Earth heats up even more. Since humans are putting more and more of the greenhouse gasses into the atmosphere, there is great concern that global warming is happening. A rise in the average temperature of the Earth of 7°F (4°C) would cause the polar ice caps to melt and the water level of the ocean to rise by up to 66 m (200 feet).

There is concern that a runaway greenhouse effect could eventually occur on Earth, as it already has on the planet Venus. There the atmosphere is almost totally CO_2 and the surface temperature is 863°F (462°C).

Hot or warm objects also emit IR rays. Hunters and soldiers use special IR cameras for night vision devices. Even when no visible light is present, these devices can detect game animals or humans.

Pit vipers are venomous snakes that use IR rays to detect prey. Located in pits near the snake's nostril, figure 3.7(a), the IR receptors can find both hot blooded and cold blooded victims, as shown in figure 3.7(b, c). Also note the slitted pupil, a characteristic of pit vipers. Non-venomous snakes have round pupils, except for members of the cobra family.

3.1.6 Microwaves

MW radiation overlaps with the low frequency end of IR; MW wavelengths extend up to about 1 cm. MW radiation has many uses; we will mention two: microwave ovens and radar. As we have seen with x-rays, electromagnetic radiation is absorbed differently by different materials. In the case of MWs, metals reflect them and other materials, such as flesh, paper and ceramics, absorb them. The vibrational frequency

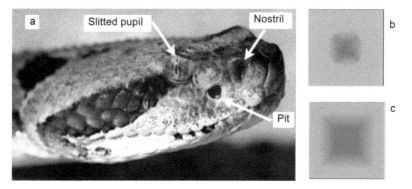

Figure 3.7. (a) The head of a pit viper. Note the rattlesnake's slitted pupil, IR detecting pit and its nostril. The pit viper's IR detectors 'see' (b) a warm blooded prey (red, rabbit) and (c) a cold blooded prey (blue, frog) against the background (pink). Photo credit: FLICKR/Clinton & Charles Robertson. CC BY-SA 2.0.

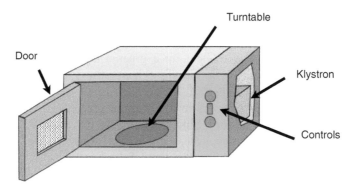

Figure 3.8. A microwave oven.

of the oxygen and hydrogen atoms in water molecules lies in the MW range of frequencies. Thus when a MW beam strikes a non-metallic substance containing moisture, the water molecules are excited into vibrating at a frequency of about 10^9 Hz. The resulting friction causes the water molecules and their surroundings to heat up. Microwave ovens use this source of heat to cook things quickly and uniformly.

The microwave oven, figure 3.8, consists of an insulated enclosure for the food, a klystron or magnetron for generating the MWs, and the controls. The window in the door is made of metal with holes smaller than the wavelength of the MWs and they bounce off, staying within the oven. However, the holes are much larger than visible light waves, and light passes right through the window so that you can watch what's cooking.

A chicken, for example, is placed on the microwave oven turntable on a ceramic dish. The turntable helps ensure that the MWs penetrate the chicken uniformly during cooking. If the chicken is not drained of liquids prior to cooking, the liquids will turn to steam and explode if confined. Cooking proceeds rapidly; the MWs penetrate the chicken completely.

Compare this to the conventional style of cooking: the chicken is placed in a preheated oven in a metal pan. The art of cookery involves being able to get the inside of the chicken cooked without: (a) burning up the outside or (b) taking so long that the meat dries out and tastes like cardboard. Placing a piece of aluminum foil over the chicken in a regular oven is not a problem, but in a microwave oven the consequences are dire: the metal foil will act as a reflector and the part of the chicken directly under the foil will not cook properly. Plus, sparks will shoot off the foil; generally not a good thing. Cooking proceeds via the slow process of heat transfer by conduction from the hot outside of the chicken to the cold insides.

All sorts of items have been put in microwave ovens, sometimes with exciting and spectacular results. One favorite is to put an old CD in a microwave oven for 10 s or less. It is strongly advisable to research in advance the consequences of any such experiment.

The reflection of MWs from metal and other substances, plus their characteristic wavelength and electronic properties, have allowed the development of radar (RAdio Detection And Ranging), figure 3.9. In this technique, pulses of MWs are sent out from a transmitter, reflected from such things as airplanes, speeding cars and falling rain, and then are received. From the timing of the pulses' reception, we can find the distance from the transmitter and the speed relative to the transmitter. At one time, the majority of long-distance telephone calls in the US traveled at least part of their way via MW links.

MW radiation also has its harmful aspects, mainly due to its ability to cook flesh. The cornea, the clear front surface of the eyeball, is very susceptible to damage by MWs. This is due partly to the cornea's lack of underlying tissue and blood vessels. Leakage of MWs from microwave ovens was at one time a serious health concern; modern construction standards for microwave ovens have relieved us of this problem.

A different health hazard due to MWs exists for users of artificial heart-muscle stimulators—pacemakers. Soon after the introduction of microwave ovens into general use, but before tightened leakage standards were imposed, microwave ovens found their way into hospital snack bars. The leakage would occasionally be experienced by a passerby wearing a pacemaker. The pacemaker would suddenly

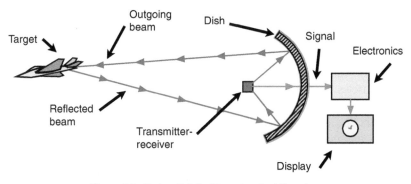

Figure 3.9. Radar: RAdio Detection And Ranging.

tell the heart to beat at MW frequencies, of 1 billion or so cycles per second. The heart would be rendered totally ineffective as a blood pump under these conditions and the unfortunate person would quickly collapse. Fortunately, on the other hand, since this happened in a hospital setting, usually there was an emergency room nearby and the person was quickly treated. This scenario is no longer much of a problem due to better MW oven construction and to the fact that pacemakers are now shielded from MWs and other forms of interference, such as that from electric razors.

3.1.7 Radio waves

Radio waves encompass the range of wavelengths from 10 cm and up. The radio spectrum, because of its extensive use in communications, is tightly regulated by the US Federal Communications Commission (FCC) in cooperation with international governmental agencies, figure 3.10. At the shortest wavelengths are the ultra-high-frequency (UHF) waves. UHF television channels 14–83 occupy this band. For example, television channel 14 has a frequency of 475.75 MHz and a wavelength of 0.63 m. The VHF (very high frequency) band contains television channels 2–13.

The entire radio spectrum is divided into specific bands for specific uses, such as cell phones, AM and FM radio, television channels, model airplane remote control devices, emergency communications, Wi-Fi, and radar. The official FCC frequency allocation document is 170 pages long.

3.1.8 Spectral response

Animals (and plants) respond differently to the regions of the electromagnetic spectrum. Figure 3.11 shows this sensitivity for a variety of organisms.

3.2 Parts of the eye

The eye is literally our window on the world. Yet, it is more than just a camera; it functions as an actual extension of the brain. As we find out more about the eye and its operation, we gain added insight and a heightened appreciation of the visual process. In this section we will look at the parts of the eye, the function of the retina and the relationship between the eye and the brain. Figure 3.12 shows a top view of the human eye. The conjunctiva is a clear, thin membrane covering the cornea and the rest of the front of the eye. The cornea merges into the sclera, the tough, white coating of most of the eyeball. This is the part of the eye that appears 'bloodshot', usually after overindulgence of one sort or another. The sclera is immortalized in the expression 'Don't fire until you see the whites of their eyes'.

A set of muscles turns the eyeball in its socket; this allows us to look in different directions without moving our heads, figure 3.13.

Light enters the eye through the cornea, a tough, transparent covering which holds in the aqueous humor. The light passes through the pupil. The iris, a sheet of circular muscular tissue, forms the pupil and controls the amount of light entering the inner portion of the eye. Because of the pigments it contains, the iris determines the eye's color: blue, gray, hazel, brown or green. Professor Hans Eiberg and his

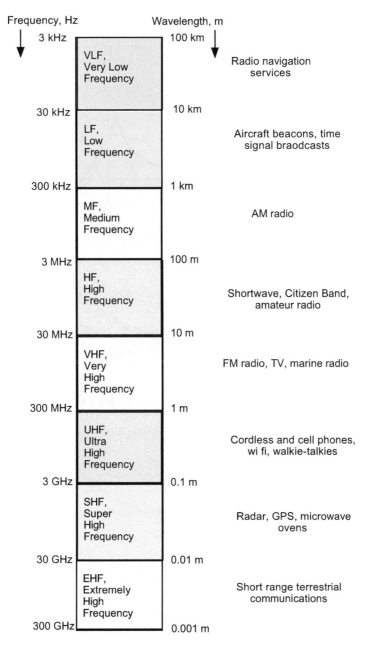

Figure 3.10. The radio frequencies for various uses.

group at the University of Copenhagen have established that all blue-eyed humans have a single common ancestor. The genetic mutation for this occurred some 6000–10 000 years ago and acted as a switch which turned off our ability to produce brown irises.

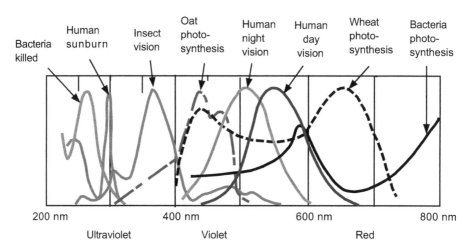

Figure 3.11. The spectral sensitivity of various plants and animals.

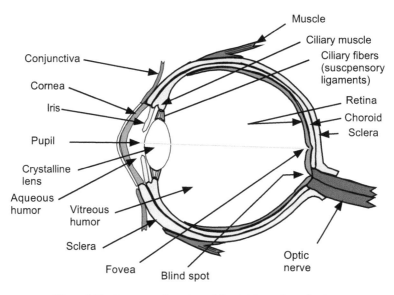

Figure 3.12. A cross-section view of the right eye, from the top.

The iris is part of the autonomic nervous system. We do not consciously control the action of the iris. In addition to reacting to the amount of light present, the iris will automatically dilate in a fight-or-flight situation and it will open when we look at something of interest. Research also tells us that we are more attracted to a person whose pupils are more expanded than otherwise. The nerve pathway for the iris dilator muscles is different than that for the iris constrictor muscles. The dilator nerves travel down the spinal column to the middle back, out of the spinal cord, back up alongside the spinal cord and in the neck along the carotid artery. From there the nerves pass into the eye through one of the ocular muscles and then, finally,

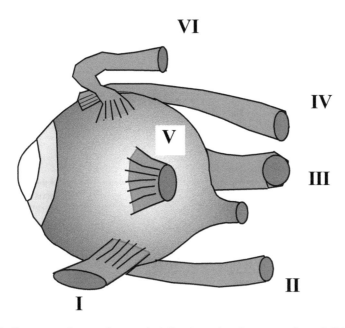

Figure 3.13. Six extra-ocular muscles, attached directly to the sclera, move the eyeball in its socket.

into the iris. The constrictor muscles, however, travel right down the optic nerve and into the eye, around its periphery and into the iris.

In earlier times, a stylish woman might carry belladonna eyedrops in her makeup kit. Belladonna ('beautiful woman' in Italian) is a muscle relaxer; application to her eyes would dilate the woman's pupils and make her appear more seductive. Interestingly, if one eye is covered and the other eye experiences a bright light, both irises will constrict.

Next the light passes through the lens and into the inner part of the eyeball. Suspensory ligaments support the lens; they are attached to the encircling ciliary muscles. When the eye is focused at infinity, the ciliary muscles are relaxed and the suspensory ligaments are tight. This causes the lens to assume a flattened shape. On the other hand, when the object to be viewed is close, the muscles contract. This causes the ligaments to loosen and the lens becomes more spherical, figure 3.14.

The lens forms the boundary between the aqueous humor, which is of a watery viscosity, and the vitreous humor, which is about the consistency of clear Jello®. In analogy with the eye as a camera, the retina is its film. Between the retina and sclera is the vascular choroid layer. The fovea, surrounded by the yellowish macula, lies directly behind the lens and pupil. As shown in figure 3.15, the retinal veins, arteries and nerve fibers all exit the eye at the optic disc or blind spot. Figure 3.16 explains how this spot can be demonstrated.

The retina itself consists of several layers, figure 3.17. Light enters from the top and passes through the ganglionic and nerve fiber layer, the bipolar cell layer, and finally into the rods and cones, the actual light sensing cells. Rods are primarily used in dim light and to detect motion. Cones are used to see fine detail and color. In the region of the macula,

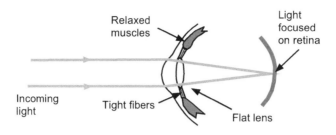

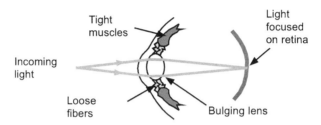

Figure 3.14. The ciliary fibers and muscles, attached directly to the sclera, work with the crystalline lens to accomplish accommodation. (a) A distant object and (b) a close object.

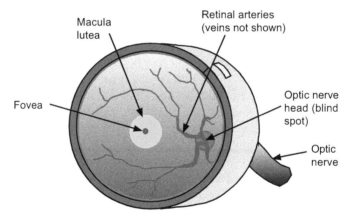

Figure 3.15. A view of the back of the inside of the eyeball, showing the fovea, macula, and blind spot.

Figure 3.16. The blind spot can be demonstrated in the following way. Close your left eye and look at the dot from about one foot away. Slowly move the page closer and farther away until the cross disappears. At this point the image of the cross is falling on your optic nerve head.

only cones are present. As one progresses outward along the retina from this region, proportionately more and more rods are found until, finally, at the periphery of the retina, there are no cones at all. The far peripheral vision can detect motion but not color. There are approximately 115 million rods and 7 million cones per eye.

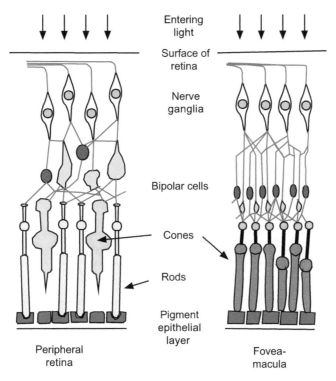

Figure 3.17. Cross-sections of two areas of the retina: the cone-rich macula and the rod-rich periphery.

To see something in dim light, we use '*averted vision*'. When we look directly at a faint object, its image falls on the macula, a region with virtually no rods. Instead we look slightly to the side of it. This casts the image onto some rods and the faint object becomes visible.

The lowest level of the retina is the pigment epithelium layer. This is the source of the 'red eyes' in color photos of human faces taken with a flash. The epithelial layer reflects any unabsorbed incident light; efficiency is increased as the rods and cones have a second chance to detect the photons. Unlike humans, animals such as frogs and dogs have a mirror layer called the *tapetum lucidum*. This layer provides increased reflectance so that in a light beam at night, the eyes of these animals appear to be bright, shining sources of light. Night-time frog hunters use this fact to locate their prey; excitement is added to the hunt because the eyes of spiders and snakes also reflect light beams!

The proportion of rods to cones is also different in man, a daylight creature, and in nocturnal animals such as owls. Figure 3.18 compares the cone/rod ratios of humans and owls.

The abundance of rods in the owl's eyes greatly enhances its night vision. Humans are primarily daytime creatures. Having more cones gives us superior bright light vision and the ability to see colors.

A photon passing through a rod or cone starts the process of vision. When the photon strikes a molecule of retinene (figure 3.19), a nerve pulse is generated.

Human

Owl

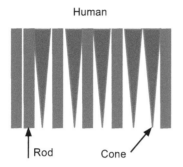
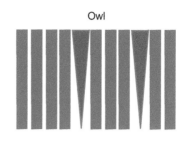

Rod   Cone

Figure 3.18. A schematic comparison of the rods and cones in the retinas of a human and an owl.

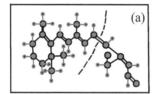
(a)
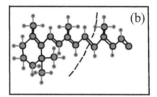
(b)

Figure 3.19. The retinene molecule (a) in darkness and (b) exposed to light. The dotted lines show which part of the retinene molecule changes its orientation when a photon hits it. This change sets off the electrochemical reaction that results in a nerve impulse. This is an example of the trigger effect: a small input of energy results in a huge release of stored up energy.

The pigmented disks stacked in the cell contain the retinene molecules, figure 3.20.

Most of the photons pass right through the stacks of disks. The few that are absorbed yield only a tiny fraction of the energy necessary for the generation of a nerve pulse. By changing the molecular arrangement of a retinene molecule, the absorbed photon initiates a trigger effect process. The rearrangement opens a hole in the retinene disk membrane and an avalanche of chemicals results.

Think of a small hole in a dam resulting in a collapse of the dam. The avalanche of chemicals in turn opens even more holes in the next disc, and so on, until finally enough energy is released to form a nerve pulse. The pulse is transmitted to the bipolar cells, then to the ganglions, and finally to the optic nerve fibers. There is a large number of interconnections among the bipolar cells and the ganglia. This results in an intelligent message being sent to the brain, as opposed to a simple television camera-like image. The wide variety of optical illusions attests to this fact.

Near the fovea, each cone has its own direct connection to a bipolar cell and ganglion. Near the periphery of the retina, there can be as many as 80 rods connected to a single bipolar cell. After 'firing off', the rod or cone must regenerate its original chemical configuration before it can detect another photon. In rods, the visual pigment is called rhodopsin, or visual purple (it's actually red). In cones there are three different pigments, each sensitive to different areas of the spectrum. As we will see, these are crucial to our ability to discern colors. The rhodopsin of the rods includes a chemical relative of vitamin A, a deficiency of which can cause 'night blindness'.

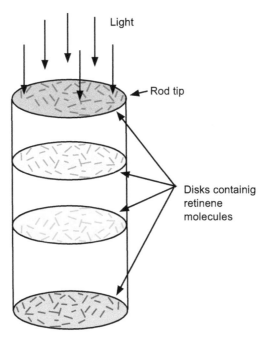

Figure 3.20. Pigment disks containing retinene (small straight lines) are stacked in the rod cells.

The visual signals in the form of nerve pulses leave the eye through the optic nerve. They pass through the optic chiasma to the lateral geniculate body and then to the visual cortex in the rear of the brain, figure 3.21. The exact details of this pathway are not known. However, it is known that the cortex sends signals back to the eye as feedback.

The eye is also connected to the autonomic nervous system. When you are physically threatened, many things automatically happen to your body. Adrenaline is released, your heart rate increases and, as mentioned before, your pupils dilate, regardless of (or in spite of) the ambient light levels. The visual cortex is also connected to other parts of the brain such as the memory regions, so that you can remember or recognize what you see. All in all, your system of vision is complicated, complex, imperfectly understood and—most important of all—amazingly effective.

The eye has to learn how to see. In a condition called 'lazy eye' or *amblyopia*, one eye does not develop its ability to 'see' even though it is physically perfect. Vision with this eye is blurry and unfocused, and the eye can even look in a different direction than the 'good eye'. In the common treatment of this condition, the eye doctor covers the 'good eye' with a patch. This forces the 'lazy eye' to learn how to see. The earlier the treatment is started the higher the success rate. After the age of about 10 years, however, the 'lazy eye' is usually incapable of recovering, even if the 'good eye' becomes completely blind.

3.3 Other eyes

Let us now take a brief look at some other approaches to vision in the animal kingdom. We will mainly be interested in the differences and similarities to our own sense of seeing.

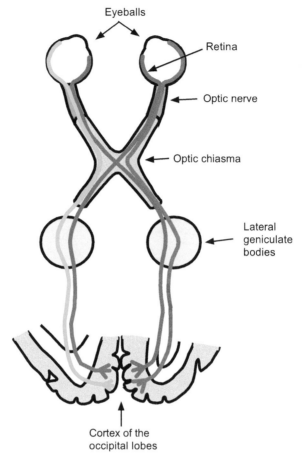

Figure 3.21. The optical pathway, simplified. After Gray, plate 722, Wikipedia Commons (http://commons. wikimedia.org/wiki/File:Gray722.png).

3.3.1 Euglena

Even some of the one-celled organisms such as the aquatic Euglena have 'eye spots' to assist them in orienting their bodies in the water. These light sensitive areas are not true eyes. They do not form images; they serve merely to detect differences in ambient light levels, figure 3.22:

3.3.2 Planaria

Planaria are small (3–15 mm, 0.1–0.6 inch), simple, multi-cellular flatworms. They glide along the bottoms of freshwater shallows throughout the world. Their two eye spots form a crude vision system and give them the nickname 'the cross-eyed worm', figure 3.23. Even though no image as such is formed, their vision system is much more sophisticated than that of the single-celled Euglena.

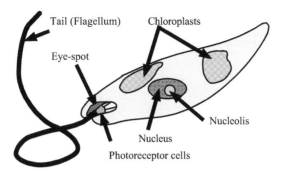

Figure 3.22. Diagram showing the eye-spot of the Euglena, a one-celled aquatic organism that can move freely (tail) and can photosynthesize light to make its own food (chloroplasts).

3.3.3 Chambered nautilus

The next step in the development of the eye is that of the chambered nautilus, a spiral-shelled relative of the squid and octopus, figure 3.24. The nautilus eye has no lens but works on the principle of the pinhole camera, figure 3.25. The light from an object is focused on the retina from any distance, without a lens. This works best if there is plenty of light.

3.3.4 Eye types

The animal kingdom displays a wide variety of eye examples, figure 3.26. Note that in all of the invertebrate eyes, the light hits the sensory cells directly. Compare and contrast this to the vertebrate eyes, where the light must travel through a layer of nerves and ganglia before striking the sensory cells. Also, the types of eyes of the jumping spider and molluscs are fixed in their heads and don't swivel. These animals compensate by having multiple eyes pointing in different directions. The scallop has over 200 eyes distributed around its rim.

3.3.5 Copilia

Perhaps the smallest animal with an organ capable of forming an image is the *Copilia*, figure 3.27. This pinhead-sized aquatic arthropod possesses eyes, each of which has a large lens and only one receptor element. Instead of having a retina, the *Copilia* has scanning muscles attached to each receptor. The muscles move the receptor back and forth, thus enabling the *Copilia* to form a crude image.

3.3.6 Insects

Insects have compound eyes, which can take on unusual forms and shapes. Figure 3.28 shows the compound eye of a mayfly. Each individual element, or ommatidium, of the compound eye has its own lens and crystalline cone. The lens and cone of each ommatidium concentrate the light onto the visual cells. These are arranged around a translucent cylinder called the rhabdome. Adjacent ommatidia

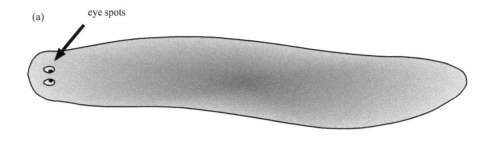

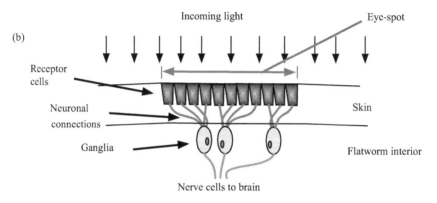

Figure 3.23. Drawing of the planaria *Dugesia subtentaculata*, a small aquatic flatworm with a simple visual system. (a) The top view of the worm shows the eye spots. (b) A diagrammatic cross-section of the visual system. The receptor cells detect incoming light. The signals go through the neuronal connections to the ganglia. Nerve cells transmit the impulses on to the planaria's brain.

are optically isolated by the pigment cells and an opaque layer. These layers also reflect unabsorbed light back up for a second chance at absorption.

The insect eye is poor in acuity, forming a crude matrix-like image, figure 3.28(e). However, it does allow the insect to detect motion extremely well, as evidenced by the ability of a housefly to escape the flyswatter so effectively.

3.3.7 Mantis shrimp eyes

Before we leave the invertebrates' visual systems and delve into the vertebrates' eyes, we should take a look at the most unusual eyes of any animal, living or fossil: those of the *mantis shrimp*. Most mantis shrimp are about the size of a human finger; however, one was caught off the coast of Florida that was 46 cm (18 in.) long. Figure 3.29 shows a peacock mantis shrimp, *Odontodactylus scyllarus*, in its approximately true colors. See figure 5.35 for labels of the parts.

The mantis shrimp's eyes are mounted on short stalks and move about constantly and independently as the mantis surveys its environment, figure 3.30.

Each mantis shrimp's eye is an example of a compound eye. A close-up of figure 3.31 shows how each eye consists of hundreds of ommatidia (figure 3.28).

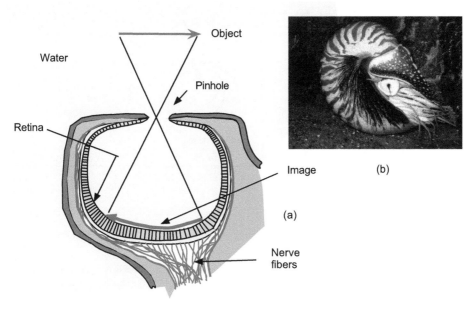

Figure 3.24. (a) The eye of the chambered nautilus. (b) The chambered nautilus. Photo credit J Baecker.

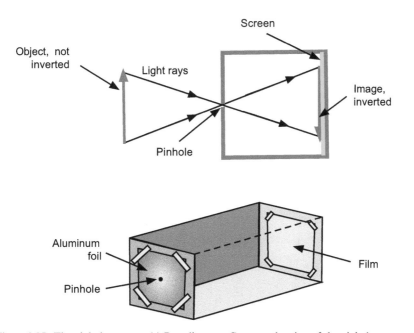

Figure 3.25. The pinhole camera. (a) Ray diagram. Cutaway drawing of the pinhole camera.

Compound eyes are excellent at motion detection; the individual ommatidia can be seen in figure 3.31(b). Note the *pseudo-pupils*, which are not pupils at all but artifacts of compound eyes. Figure 3.32 shows the compound eye of a praying mantis with its pronounced pseudo-pupils.

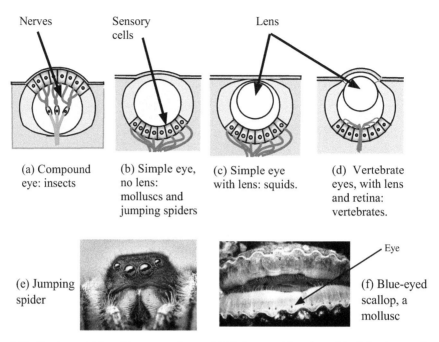

Figure 3.26. Eye types: (a) Insects' compound eyes. (b) Molluscs and jumping spiders' simple eyes. Lenses in (c) squids' eyes and (d) vertebrates' eyes. (e) Jumping spider. This Phidippus audax image has been obtained by the author from the Wikimedia website where it was made available by Opoterser under a CC BY-SA 3.0 licence. It is included within this article on that basis. It is attributed to Opoterser. (f) Blue-eyed scallop, a mollusc, the image has been obtained by the author from the Wikimedia website https://en.wikipedia.org/wiki/File:Scallop_eyes.jpg, where it is stated to have been released into the public domain. It is included within this article on that basis.

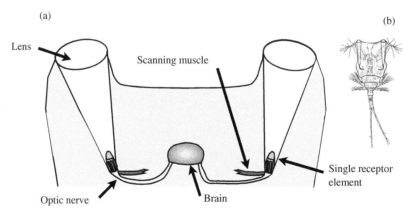

Figure 3.27. (a) Diagram of the head of a *Copilia quadrata*, a small aquatic arthropod. The scanning muscle moves the single receptor back and forth, forming a crude image. (b) A female *Copilia*, around 3 mm (0.1 inch) long. (From figure 53, p 153, Calman W T 1911 *The Life of Crustacea* (London: Methuen & Co. Ltd).)

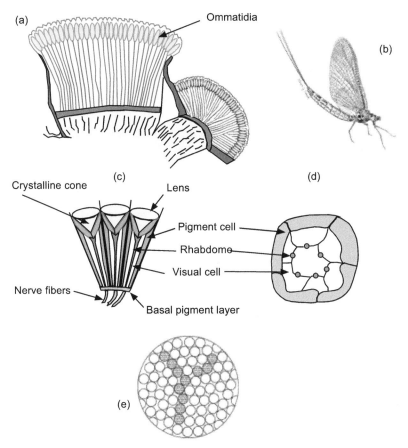

Figure 3.28. Insect eyes. (a) The compound eye of the mayfly *Cloeon dipterum*. The large part at the top is for detailed vision and the smaller unit to the right gives wide-angled, coarse vision. The eye is made up of many individual ommatidia. (b) Picture of a mayfly, illustration by Karen J Couch. (c) Shows three ommatidia and their parts, including the rhabdomes. (d) Depicts the cross-section of a single ommatidium. The visual cells are each distinct and some are tuned to different wavelengths. (e) Shows the type of mosaic image that, for example, an insect might perceive when viewing the letter 'Y'.

The mantis shrimp midband, figure 3.31(a), has six parallel rows of ommatidia sensors.

The first four rows have 16 types of color sensors capable of being individually tuned. For comparison, non color-blind humans have three types of color sensors and can perceive 10 million different colors. Just imagine what the mantis shrimp sees with 16! The last two rows of the midband are devoted to the two types of polarized light: circular and linear. See figure 4.66 for a description of how polarized sunglasses work using linear polarization.

The mantis shrimp is the only animal known to be able to detect circularly polarized light, too. Recent advanced optical devices, such as CD and DVD players,

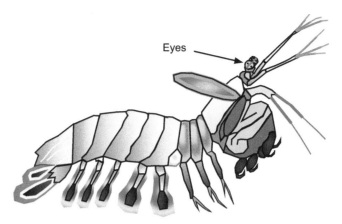

Eyes

Figure 3.29. Drawing of the peacock mantis shrimp, *Odontodactylus scyllarus*. The coloring is close to realistic; the parts are labeled in figure 5.35. Mantis shrimps are about 13 cm (5 in.) and are also called 'thumb splitters'.

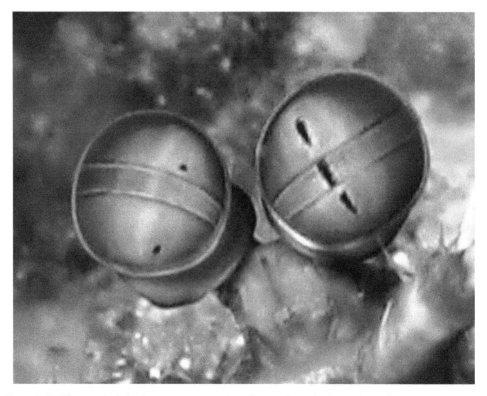

Figure 3.30. The mantis shrimp's eyes are mounted on short stalks. NSF image: https://commons.wikimedia. org/wiki/File:Mantis_Shrimp_Eyes.JPG. This image has been obtained by the author from the Wikimedia website where it was made available by Alexander Vasenin under a CC BY-SA 3.0 licence. It is included within this article on that basis. It is attributed to Alexander Vasenin.

(a)

(b)

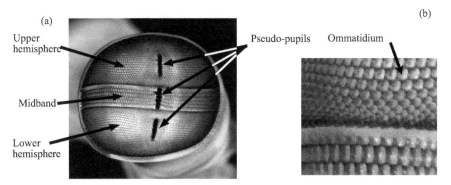

Upper hemisphere

Midband

Lower hemisphere

Pseudo-pupils

Ommatidium

Figure 3.31. Photos of a mantis shrimp eye. (a) Showing the upper and lower hemispheres, the midband, and the pseudo-pupils. (b) Detail showing individual ommatidia. Photos credit Roy L. Caldwell/University of California, Berkeley). NSF https://nsf.gov/discoveries/disc_images.jsp?cntn_id=299671&org=NSF.

Pseudo-pupils

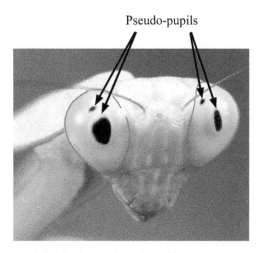

Figure 3.32. Preying mantis head showing its pseudo-pupils and its compound eyes, outlined in red for clarity. Rhombodera_basalis_2_Luc_Viatour_cropped.jpg. This figure has been adapted by the author from a Rhombodera basalis image obtained from the Wikimedia website. This image was made available by Luc Viatour under a CC BY-SA 3.0 licence, and is included within this article on that basis. It is attributed to Luc Viatour.

use what is called quarter-wave plates to detect circularly polarized light in at most a few colors. The mantis shrimp vision system works across the entire visual spectrum! The two hemispheres of the mantis shrimp are good at detecting green light and UV light, and in recognizing forms and motion. The mantis shrimp can see objects with all three parts of the same eye for depth perception, an ability called trinocular vision. Using the other eye, too, the mantis shrimp has an unparalleled degree of depth perception.

Finally, as well as generating images, the mantis shrimp's vision system is capable of 'spectral tuning', a unique ability. With it, the mantis shrimp can adjust its sensitivity to, say, the red range of wavelengths, so that it can quickly detect red changes in its environment. The mantis shrimp can also shorten its image processing time and speed up the decision-making process, improving its survivability. A good thing, too: the mantis shrimp has a large number of predators, including humans, who find them delicious when properly cooked. Numerous recipes are online. Live mantis shrimp can be purchased for aquaria use. Beware, however, that they have been known to break aquarium glass and that their nickname is the 'thumb splitter'.

3.3.8 Largest eye sizes

Animal eyeballs come in a variety of sizes. To consider the largest animal eyes, categorization is in order: land or aquatic, living or extinct and eye size in relation to body size. For example, the mammal with the largest eyes in comparison to its body size is the tarsier, figure 3.33(b), which weighs about the same as a baseball or two D cell batteries (between 80 and 160 g (2.8–5.6 oz)). For comparison, the eye of a giant squid, figure 3.34 is about the size of a basketball and has a mass of as much as 10 kg (22 pounds), 125 times higher than the tarsier's entire body mass!

The giant squid has the largest eyes of any living animal, aquatic or land; but the extinct dinosaur (not a fish) ichthyosaur, *Ophthalmosaurus*, figure 3.34b, had even larger eyes. The ichthyosaur lived 200 million years ago and the giant squid is alive (and well) today. Yet the eyes of both had the same purpose: to see in the almost complete darkness of the ocean deeps. The sperm whale, on the other hand, uses a sophisticated and powerful sonar system to locate its prey in the same depths, chapter 5.

Back on land, the ostrich, figure 3.35(a), has the largest eyes of any living *land* animal, about 5 cm, or 2 in. in diameter, bigger than its brain. The extinct Giganotosaurus, figure 3.35(b), slightly larger than the Tyrannosaurs Rex, surpassed this with eyes of twice this size. In comparison, human eyes have a diameter of 3.8 cm (1.5 in.) in size.

3.3.9 The nictitating membrane

The *nictitating membrane*, sometimes called the third eyelid, is a tough covering that fits between the eyelid and cornea of many animals, including sharks, raptors, frogs, and humans. Right before a raptor's impact with its prey, the nictitating membrane

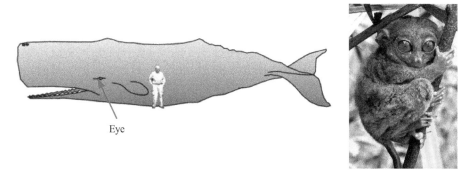

Eye

Figure 3.33. In comparison to its body, the sperm whale (a) has tiny eyes whereas the tarsier (b) has the largest eyes of any mammal, when body size is considered. This Tarsier image has been obtained by the author from the Wikimedia website where it was made available under a CC BY-SA 2.0 licence. It is attributed to mtoz.

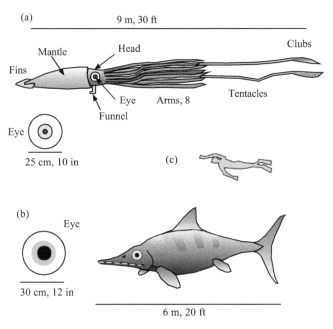

Figure 3.34. Comparative eye sizes. (a) Giant squid, largest eyes of any living creature. (b) The extinct dinosaur ichthyosaur, *Ophthalmosaurus*, extinct, largest eyes ever. (c) Human, approximately to scale.

Figure 3.35. (a) Ostriches have the largest eyes of any living land animal. Photo credit: William Warby. (b) The dinosaur *Giganotosaurus*, which was slightly bigger than Tyrannosaurs Rex, had the largest eyes of any land animal, living or extinct. This image has been obtained by the author from the Wikimedia website where it was made available by Durbed under a CC BY-SA 3.0 licence. It is included within this article on that basis. It is attributed to Durbed.

quickly moves across the cornea and protects it, figure 3.36. Figure 3.37 shows how you can see your own nictitating membrane.

3.3.10 Rotating pupils

Horses, cows, goats and sheep are prey animals. Their horizontal pupils give them an enhanced wide-angle view in the horizontal plane, all the better to detect predators with. When they lean down to drink or eat, their ocular muscles rotate

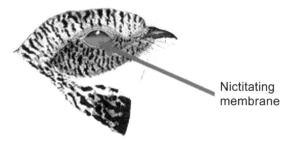

Figure 3.36. The partially closed nictitating membrane of the common poorwill (or nightjar), shown in pink for emphasis. Public domain art.

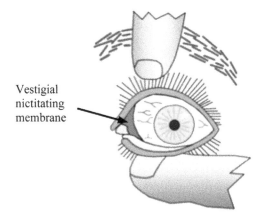

Figure 3.37. The human vestigial nictitating membrane revealed; enhanced color for emphasis.

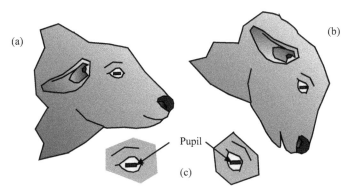

Figure 3.38. Drawing of a sheep's head showing the pupil (a). When the sheep lowers its head to drink or graze, the eyeball rotates, keeping the pupil parallel to the ground (b). This helps the sheep detect and thus avoid wolves and other predators. (c) Pupil detail.

their eyeballs to keep the pupils horizontal, figure 3.38. Your own eyeballs also rotate in their sockets giving you the ability to look in different directions without moving your head. 'Rolling your eyes' also lets you communicate non-verbally on the possibly ridiculous statements or actions of others.

This brings us to the end of chapter 3, The Eye. Next we will discuss some interesting aspects of waves in chapter 4.

IOP Publishing

The Everyday Physics of Hearing and Vision (Second Edition)

Benjamin de Mayo

Chapter 4

Wave properties

Waves come in many forms; light is one form of wave and sound is another. Of the many wave forms, all have certain things in common. In this chapter, we will examine some of these common phenomena and we will see that their effects on us range from the profound to the mundane.

Whenever a wave strikes a boundary, a combination of things occur, figure 4.1. For convenience, we show the wave as a straight line, or ray. The incident ray traveling in medium 1 hits the surface. The *normal* line is drawn perpendicular to the surface at the point where the ray strikes the surface. Part of the wave is reflected and part enters the medium 2. We call the part in medium 2 the refracted ray; note that its direction has changed. This refracted part then passes through the medium without bending until it strikes the second boundary. Part of it leaves medium 2; we call this the transmitted ray. The transmitted ray changes direction as it leaves medium 2. A portion reflects back and stays in medium 2. The transmitted ray continues until it strikes another boundary.

Some materials will almost completely absorb waves hitting them. We call these substances opaque. Other materials transmit most of an incident wave and are called transparent or translucent. For example, plate glass is transparent since you can see images through it, but frosted glass is translucent. Even though light is transmitted, the images are blurred by the frosting on the surface of the glass.

4.1 Reflection

The wave process called reflection is very important to our senses of sight and hearing. In reflection, a simple relationship exists between the incident and reflected rays. Called the law of reflection, it says that the angle between the normal and the incident ray is equal to the angle between the normal and the reflected ray, figure 4.2. For this to work, the reflective surface must be smooth and flat.

We see that for whatever angle the wave strikes the boundary, the reflected wave bounces off at the same angle. Please note that reflection occurs for all types of

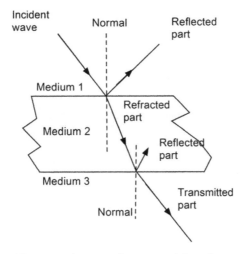

Figure 4.1. A wave strikes a smooth boundary.

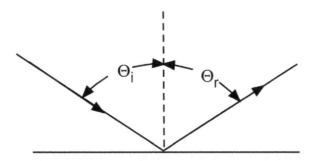

Figure 4.2. Reflection off a smooth flat surface: the angle of incidence, Θ_i, is equal to the angle of reflection, Θ_r.

waves. For each type of wave, different materials may be excellent or poor reflectors, however. For example, smooth wood makes an excellent sound reflector, but it would be risky to use it as a mirror for shaving. We will see shortly what the criteria are for good mirrors.

There are basically two types of reflection: (1) *regular* or '*specular*' and (2) *diffuse*; figure 4.3. In specular or regular reflection, a well-defined beam of waves strikes a 'smooth' surface and, without losing its definition, bounces off the surface. On the other hand, in diffuse reflection, a well-defined beam is broken up and leaves the non-smooth surface in many directions. The inset of figure 4.3(b) shows that at each point of impact, regular reflection does in fact occur. However, since the surface is

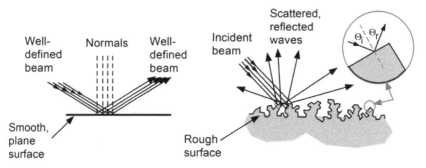

Figure 4.3. A well-defined beam of parallel rays strikes (a) a smooth, flat surface and (b) a rough surface. Even though the law of reflection holds on a microscopic scale (inset), the rough surface breaks up the beam.

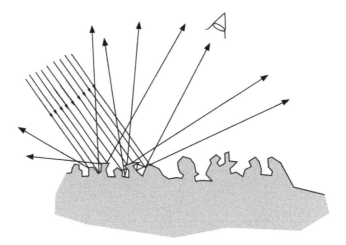

Figure 4.4. Diffuse reflection makes possible the visibility of an object from many directions.

rough, the angles of incidence and reflection vary drastically across the surface. Even though the law of reflection holds on a microscopic scale (inset), the rough surface breaks up the beam.

Diffuse reflection is extremely important in vision. Of the objects that we see, only a few emit their own light. Most objects, such as ourselves, depend on reflected light to be visible. If the reflection off objects were totally specular, what we would see is a number of very bright, sharp points of light, similar to sunlight reflected off car windows. Figure 4.4 shows how diffuse reflection allows an object which is illuminated from one direction to be visible from a number of directions. Smoothness is an important necessity for a mirror, which may be defined as a device for specularly reflecting waves. The surface of the object must be smooth to the order of about one wavelength for the surface to be acceptable as a mirror. A high ratio of reflection to absorption is also advantageous. Even though they were highly desirable for cosmetic purposes, mirrors were difficult for the ancients to construct. A basin of standing water possesses a smooth surface, but the absorption of light is rather high. However, we can polish metal to be smooth and flat. Metal is also quite reflective to visible light. Figure 4.5 shows the construction of modern mirrors. We attach a thin layer of metal directly to the back of a sheet of glass. A backing is added for strength and protection against scratching.

The image produced by a plane mirror is 'perverted'. As shown in figure 4.6, when you look at yourself in a mirror with your right arm raised the person looking back at you appears to have their *left* arm raised. However, even though right and left are seemingly reversed, up and down are not (such an image would be inverted).

Consider now a highly reflecting surface that is curved and smooth. Using rays we can see what happens to a wave which hits such a surface. First, let's look at a concave or bowed-in surface, figure 4.7. This concave mirror is a concentrating lens; if a wave approaches so that the rays are parallel, then the energy of the wave is concentrated at the *focal point*. Note that, as the inset shows, at each point on the surface of the mirror the law of reflection holds. Such lenses are used in a number of ways: solar concentrators built this way allow food to be cooked at the focal point. Many optical and radio telescopes for astronomy are based on the use of a reflecting lens. Isaac Newton invented this type of instrument, figure 4.8. The Hale Telescope at the Palomar Observatory, one of the largest in the world, has a mirror 5 m (16 feet) in diameter.

We can pick up faint sounds by placing a microphone at the focal point of a reflecting lens. Television sports casts do this to detect quarterbacks' verbal commands and the sounds of physical contact. Many satellite dishes and radio telescopes, figure 4.9, have concave reflecting mirrors. The radio telescope at Arecibo in Puerto Rico has a mirror diameter of 300 m (1000 feet). The first such radio telescope was made of chicken wire. Even though this would not seem to be smooth enough to serve as a mirror, the fact that radio waves have so much longer a wavelength than visible light means that the 1 cm (1/2 in.) openings of chicken wire are quite smooth by radio wave standards.

A convex mirror forms a diverging lens, figure 4.10. Note that the law of reflection holds at each point on the surface, as the inset shows. We also see a sort of focal point behind the lens from which the rays seem to diverge. These lenses are found in fun-houses at amusement parks and where peering around a corner is useful, such as at hospital corridor intersections. The diverging reflecting lens differs

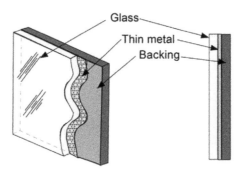

Figure 4.5. Mirror construction details. Glass supports the metal film, which does the actual reflecting. The backing serves to prevent scratching of the thin metal film.

Figure 4.6. A perverted image. Left and right are reversed, but not up and down. The image has a left arm raised whereas the object has a right arm up.

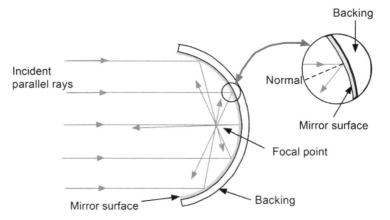

Figure 4.7. A cross-section of a parabolic concave mirror: a concentrating lens. The inset shows that at each point on the surface of the mirror, the law of reflection occurs.

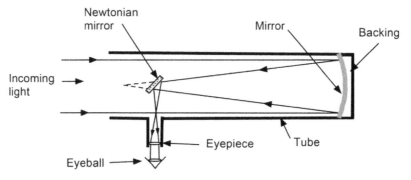

Figure 4.8. The reflecting telescope, invented by Isaac Newton.

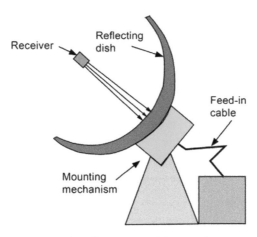

Figure 4.9. A radio telescope of the steerable type.

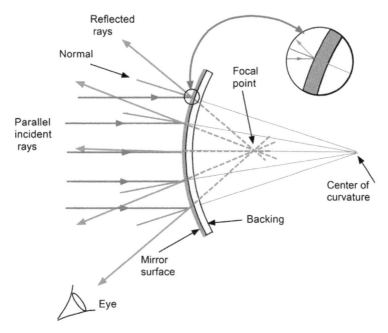

Figure 4.10. A cross-section of a convex mirror: a diverging lens. As the inset shows, the law of reflection holds at each point on the mirror's surface.

from diffuse reflection in that if we put our eye at the position shown in figure 4.10, we still see the image of the wave source. Such is not the case for diffuse reflection; compare figures 4.4 and 4.10. The diffuse reflector of figure 4.4 may be called a 'non-imaging diverging reflector'.

The reason for mentioning non-imaging devices is that our ears make important use of a 'non-imaging concentrating reflector' in the form of the auditory canal; figure 4.11 shows this schematically. The sound wave energy is concentrated as it

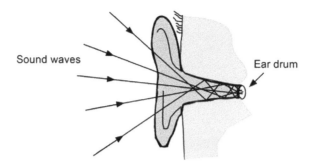

Figure 4.11. The outer ear and auditory canal concentrate sound waves by using reflection.

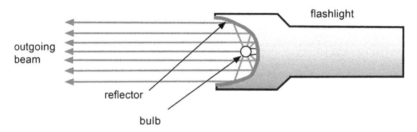

Figure 4.12. Flashlights and auto headlights use reflection to form a concentrated, parallel beam of light.

progresses down the canal to the ear drum. The outer ear assists in this to some degree, although we can tell from birds that the outer ear is not indispensable (figure 2.14). The earliest hearing aids were large horn-like devices which performed in the same manner, but on a larger scale, figure 2.20.

We can use concentrating reflectors in reverse, too. The flashlight, figure 4.12, and megaphone both use reflection to obtain the desired results: a narrowed beam of radiation.

4.2 Absorption

The opposite of reflection is absorption. Under many circumstances absorption is desirable; sometimes it can even be enhanced by using reflection. Figure 4.13 shows how this is achieved in acoustically absorbent ceiling tiles. The tiles are designed to soak up sound waves. By clever construction, the number of reflections is increased by having many small holes in the tiles. Before a sound wave can escape from one of the holes, it will undergo many reflections. Part of the energy of the sound wave is absorbed each time it strikes the material. The amount of sound that finally exits each hole is very small indeed.

4.3 Refraction

Refraction is the bending of the direction of a wave as it travels from one medium into a second medium. The wave's direction is changed because the speed of the wave is different in the two media. As in reflection, all types of waves refract.

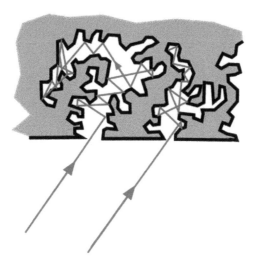

Figure 4.13. In an acoustic ceiling tile, absorption is maximized by increasing the number of reflections undergone.

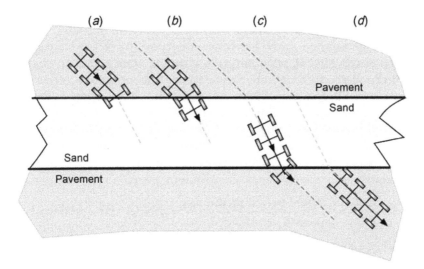

Figure 4.14. An octocycle, a new type of eight-wheeled dune buggy, goes from a pavement into sand and back onto the pavement. Notice how in (a) the right front tire is pulled to the right as it enters the sand and slows down. This process continues (b) with the result that the direction of the octocycle is changed. Once the octocycle is completely in the sand, its direction does not change until (c) as it leaves the sand. Then the right front wheel speeds up and the octocycle resumes its original direction (d).

How does a change in speed change the direction of the wave? Consider figure 4.14, which shows the top view of an eight-wheeled vehicle (an 'octocycle') going from a paved surface into sand. The wheels travel much more easily on the pavement than they do on sand, so that when the first wheel hits the sand, it slows down and pulls to

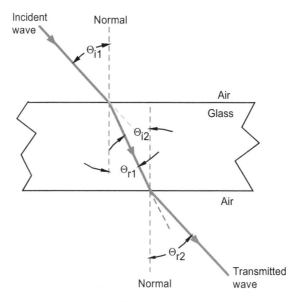

Figure 4.15. Refraction of a light beam passing through a parallel plate of glass.

the right (a). This process continues with each wheel until finally, the vehicle is completely in the sand and now has a new direction (b). What happens when the process is reversed? That is, what happens when the octocycle goes from sand (the medium of slower speed) back onto the pavement (the medium of higher speed)? As shown in figure 4.14, now the wheels pull to the left when they reach the pavement.

Figure 4.15 shows this situation using rays. When a wave goes into a medium in which its speed is slower, it bends *toward the normal*. The angle of Θ_i is greater than the angle of refraction Θ_r. When the ray enters a medium in which it speeds up, it bends away from the normal and Θ_i is less than Θ_r. We can express this mathematically:

$$\frac{\sin \Theta_i}{\sin \Theta_r} = \frac{\text{speed in first medium}}{\text{speed in second medium}} = n(\text{index of refraction}).$$

Since the speed of light in glass is slower than in air, the angle of incidence Θ_{i1} is *larger* than the angle of refraction Θ_{r1}: the refracted ray bends *toward* the normal when the wave enters a medium in which its speed is slower. When the wave leaves the glass and enters the air again, it speeds up and the angle of incidence Θ_{i2} is *smaller* than the angle of refraction Θ_{r2}: the ray bends *away from* the normal. Note that the ray leaves the glass at the same angle that it entered: $\Theta_{i1} = \Theta_{r2}$.

This relation is known as *Snell's law*, or the *law of refraction*, after Willebrord Snellius (1591–1626), a Dutch mathematician and astronomer. The angles Θ_i and Θr are numbers between 0° and 90° and the sines of these angles vary between 0 (for 0°) and 1 (for 90°).

Thus, we can use Snell's law to find the incident angle Θ_i if the refracted angle Θ_r and the index of refraction n are known. Furthermore, the largest angle of refraction occurs for the largest angle of incidence. And finally, the difference between Θ_i and Θ_r will be the largest when n, the index of refraction, is the largest; that is, when the difference in the speed in each medium is the largest. Figure 4.16 shows several cases for a beam of light. Lateral displacement of a wave is achieved by using a medium with parallel sides, such as a piece of glass, figure 4.17.

The speed of the wave changes upon refraction, but what about the wavelength and frequency? A beam of light enters a glass plate at a 90° angle and slows down. The beam travels straight on and then passes back out of the glass, figure 4.18. From Chapter 1, $w \times f =$ a constant, the speed of the wave, where w is the wavelength and f is the frequency. Furthermore, the frequency f must stay the same, otherwise there would be gaps or bunches in the wave leaving the glass. If the speed drops and the

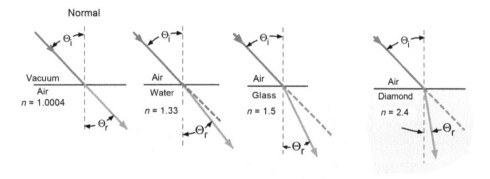

Figure 4.16. A beam of light refracts into four different media.

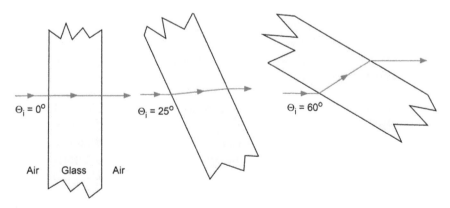

Figure 4.17. Lateral displacement of a light wave occurs when it passes through a parallel plate of glass. Different angles of incidence Θ_i result in different amounts of displacement.

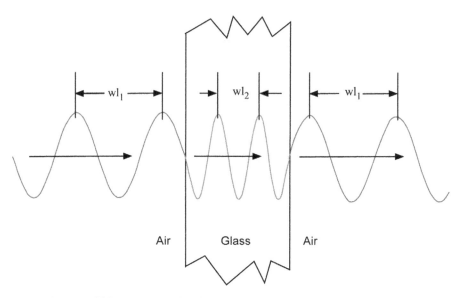

Figure 4.18. A wave which enters a medium in which it travels slower. Its wavelength w is reduced. Its frequency remains the same.

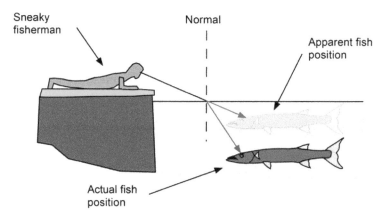

Figure 4.19. The sneaky fisherman not only sees the fish at the wrong position (shallower than reality), but he is also still visible to the fish.

frequency stays the same, the wavelength w will drop, too. Figure 4.18 shows a shorter wavelength in the glass.

Refraction has some interesting features. The sneaky fisherman in figure 4.19 thinks that by lying down on the bank of the pond, he can approach the fish undetected. However, the fish can still see the fisherman, due to refraction. The fisherman also observes the fish to be at a lesser depth than he actually is. Another feature of refraction is related to the fact that, at an interface between two media, reflection and refraction both occur and, as shown in figure 4.20, the angle of

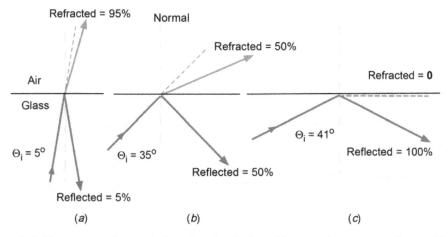

Figure 4.20. The amounts of energy in the reflected and refracted beams varies with the angle of incidence.

incidence in a large part determines the proportions of reflected and refracted energy.

4.3.1 Total internal reflection

Now consider what happens when the wave goes from a region in which its speed is low into one in which its speed is higher, for example, as in light going from glass into air. Figure 4.21 shows that as the angle of incidence increases, the refracted ray bends further and further from the normal line. Also, more and more of the energy is in the reflected beam. For larger angles as well there is no refracted beam and 100% of the energy of the wave is found in the reflected wave. This is called *total internal reflection* and it can occur for any type of wave, including water waves and sound waves. The critical angle depends on the index of refraction n, which is the ratio of the speeds of the beam in the different media. For total internal reflection, the index of refraction is less than one, a consequence of the wave going from glass (slow) to air (fast).

Figure 4.22 lists the index of refraction n and the critical angle Θ_c for light going from various materials into air.

Total internal reflection has some interesting applications. For example, in diamonds the low critical angle, combined with proper faceting, results in a jewel that sparkles. The diamond is cut so that although much of the light falling on it is absorbed, the light can leave only along certain directions—it thus 'sparkles'. Because diamond's critical angle is so low (24°), any light ray inside the diamond must hit a facet at 24° or less if it is to escape, figure 4.23. The light therefore enters from any angle but can leave only along certain ones. This focusing effect produces the sparkle.

When you purchase a diamond, you are buying three things (physically speaking): a low critical angle, a proper cut and hardness. As figure 4.22 shows, some artificially produced materials have even lower critical angles than diamond. However, since they are not nearly as hard as diamonds, they are easily scratched. They are also

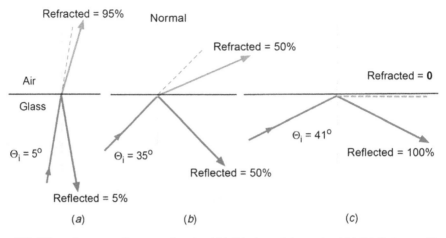

Figure 4.21. When a beam goes from a medium in which it is slower into one in which it is faster, *total internal reflection* can occur. For any angle of incidence Θ_i greater than the critical angle Θ_c (41° for glass to air), 100% of the incident light is reflected back into the glass and none of it leaves the glass. For (a) and (b) the angle of incidence is lower than the critical angle Θ_c. (c) One can see what happens when the angle of incidence equals the critical angle, 41° for glass to air: the refracted beam disappears completely.

MATERIAL	CRITICAL ANGLE (DEGREES)	INDEX OF REFRACTION
Ice	49.8	1.310
Water	48.6	1.333
Fused Quartz (SiO2)	43.3	1.458
Crown Glass	40.5	1.541
Flint Glass	31.9	1.890
Calcite (CaCO2)	37.0	1.658
Diamond	24.4	2.417
Strontium Titanate (SrTiO3)	24.4	2.417
Rutile	22.4	2.621
Air (relative to vacuum)	88.7	1.000276

Figure 4.22. The critical angle of light going from various materials into air and for light going from air into a vacuum, plus the index of refraction for each case (using yellow light (wavelength = 5.8×10^{-7} m)).

much less expensive than the real thing. A 'fake' diamond ring can be a bargain—if one remembers to take off the ring before doing manual labor.

Light pipes are constructed so that total internal reflection occurs at the surface boundary, which in turn reduces loss of the light to a minimum, figure 4.24. The light is funneled along the pipe like water in a hose. Each time the light hits the inside of the light pipe, total internal reflection occurs and no light escapes the light pipe.

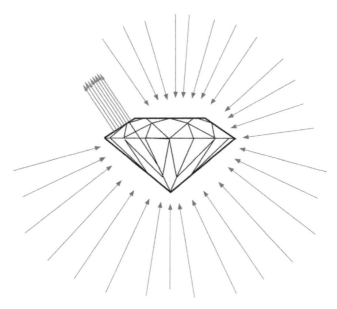

Figure 4.23. The diamond collects light from all angles but, because of total internal reflection and its low critical angle, the diamond concentrates and re-emits the light only in narrow cones centered on the facets—hence the sparkle.

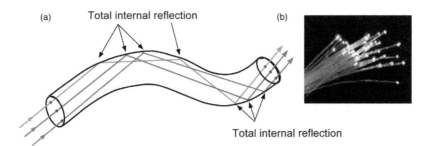

Figure 4.24. (a) A light pipe. (b) A fiber optics bundle. Total internal reflection minimizes the loss of light from the beam. Photo credit: http://commons.wikimedia.org/wiki/Optical_fiber#mediaviewer/File:Fibreoptic.jpg. This image of optical fibres has been obtained by the author from the Wikimedia website where it was made available under a CC BY-SA 3.0 licence. It is attributed to BigRiz.

Fiber optics devices use a large number of small diameter pipes, each coated with a material of suitable refractive index. These are joined together in a flexible fiber optics bundle.

Besides being used extensively in communications, physicians use fiber optics as double light pipe devices called endoscopes or cystoscopes. The instrument is inserted into the body; one fiber bundle serves to transmit light to the place of interest, the other sends an image back to the physician's eye, as shown in figure 4.25.

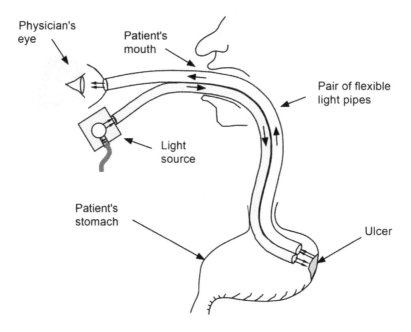

Figure 4.25. Fiber optics in medical use—an endoscope is used to detect an ulcer.

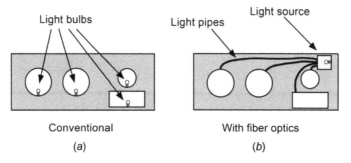

Figure 4.26. (a) A conventional automobile illuminated dashboard instrument panel: four bulbs are needed. (b) Fiber optics illumination: one bulb, four fiber optics bundles. LEDs are used now.

The automobile industry has investigated the use of fiber optics bundles to replace all of the individual dashboard light bulbs with a single bulb, figure 4.26. Nowadays, light emitting diodes (LEDs) are used.

Total internal reflection can produce interesting effects in water fountains which are illuminated by underwater lights (figure 4.27).

Binoculars are significantly shortened by using two 45° prisms; total internal reflection reduces light-loss in the prisms (figure 4.28). The prisms also present the image right side up and non-inverted.

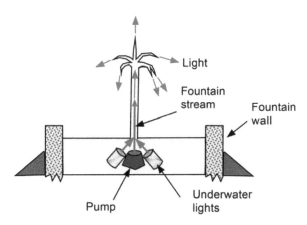

Figure 4.27. Using fountain streams as light pipes. Light is transmitted up the water column; light intensity losses are minimized since total internal reflection occurs as the light travels up the column.

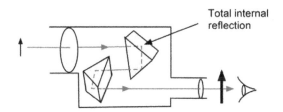

Figure 4.28. Compact binoculars use prisms and total internal reflection.

4.3.2 Refracting lenses

We make refracting substances into lenses that either converge or diverge. Consider, for example, parallel rays of light striking a double convex piece of glass, figure 4.29. All of the wave energy converges at the focal point, making this a converging lens. Notice that the rays change direction only at the surface of the lens. Once inside the lens, the ray travels in a straight line through the lens to the outer surface.

We can also make *diverging lenses*, figure 4.30. The double concave surface of the lens causes the incident wave to spread out from an imaginary 'focal point'. Again, the wave's direction changes only at the interface of the lens and the surrounding air.

Before addressing the uses of refracting lenses, let us mention briefly *Fresnel lenses*. Georges-Louis Leclerc (1707–1788) invented this type of lens and in 1820, Augustin-Jean Fresnel (1788–1827) first used them in lighthouses. A chain of coastal lighthouses warned ships that the coast was near. As figure 4.31 shows, a center portion of the lens is removed, leaving only the curved surfaces in concentric rings.

This was a great benefit to lens makers. Now a lighthouse concentrating lens could be made larger and lighter. The lighthouse beam would be much brighter and more concentrated, and the beacon could be seen from farther away.

Plastic Fresnel lens are now widely used in overhead projectors, cameras, and solar collectors, figure 4.32. You can find Fresnel lens magnifiers at your local drug

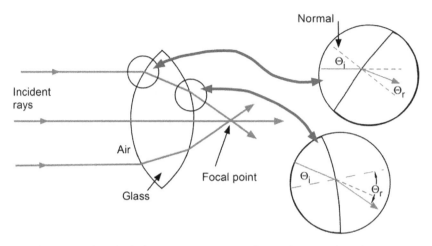

Figure 4.29. The cross-section of a double convex converging refracting lens. Parallel incident rays are brought to a focus at the focal point. The insets show how refraction occurs only at the surfaces.

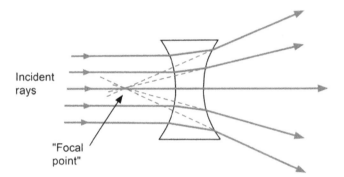

Figure 4.30. A cross-section of a double concave refracting diverging lens. Parallel incident waves are spread out and appear to originate at the 'focal point'.

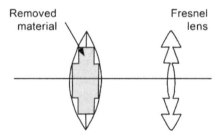

Figure 4.31. Cross-sections showing how a Fresnel lens consists of the curved surfaces of the lens, with most of the superfluous material discarded.

store or pharmacy in the senior citizen section (look for the area containing canes, walkers and crutches).

Refracting lenses are used in a multitude of optical devices, not the least of which are our own eyes. As shown in figure 4.33, the human eye has two refractors, the lens

Figure 4.32. An example of a plastic, mass produced Fresnel lens.

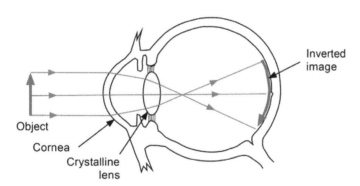

Figure 4.33. Light entering the human eye is refracted first by the cornea and then by the crystalline lens. This results in an inverted image being formed on the retina.

(naturally) and also the cornea. The light is focused on the retina, a process known as *accommodation*.

Since the amount of refraction depends on the relative difference in the speed of the wave in the two media of interest, seeing underwater is very different from seeing in air. This is because the speeds of light in water and in the cornea are very similar. This greatly reduces the refractive ability of the cornea for the aquatic observer. For proper underwater vision, divers can purchase special contact lenses that fit directly over the cornea. An effective alternative is a facemask; it allows air to be at the surface of the cornea, thus restoring its refractive power. Fish and other underwater animals have especially thick lenses in their eyes to compensate for the relative lack of corneal refraction, figure 4.34.

As shown in figure 4.35, lenses can correct for either near-sighted vision (*myopia*), in which the eye is incapable of bringing the image of a distant object to focus on the retina, or for far-sighted vision (*hyperopia*) in which a nearby object causes difficulty.

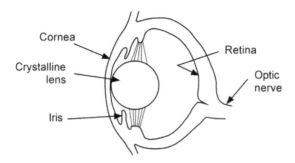

Figure 4.34. The fish eye has an almost spherical lens.

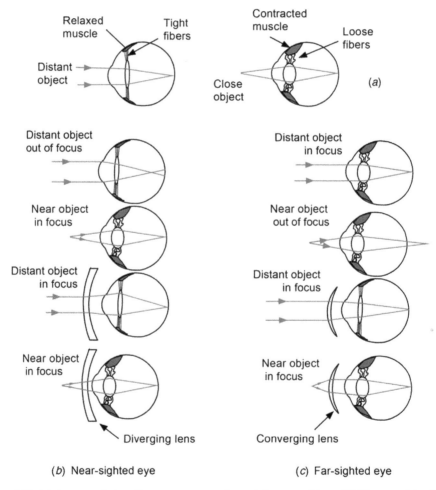

(*b*) Near-sighted eye   (*c*) Far-sighted eye

Figure 4.35. (a) Accommodation allows us to see nearby and far away objects. (b) The near-sighted eye is corrected with a diverging lens. (c) The far-sighted eye needs a converging lens for correction.

Figure 4.36. An astigmatism test chart. Remove your glasses and hold it at arm's length. Look at it with one eye; if any of the lines look darker, then you have an astigmatic condition in that eye. Not to worry; astigmatism is not only common but also correctable with contacts or regular eye glasses.

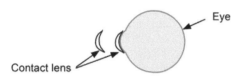

Figure 4.37. A contact lens is placed directly onto the cornea.

Astigmatism, a condition of uneven curvature of the cornea or lens, can also be corrected, figure 4.36. Almost two thirds of all people have some degree of astigmatism. Contact lenses offer cosmetic, safety, and peripheral vision benefits, figure 4.37. With advancing age, a condition known as *presbyopia* develops; the lens becomes less and less flexible, making accommodation more and more difficult. Then bifocals, glasses whose top portion aids near-sightedness and whose bottom portion aids far-sightedness, are sometimes helpful, figure 4.38.

The magnifying glass, a useful device for making an object appear larger, is made using a double convex refractor, figure 4.39. When an object is closer to the lens than focal point 2, a magnified, non-inverted image will result. Magnification with lenses may be an old process. In 1885, archaeologists found a piece of clear, convex quartz in Assyrian ruins dating back to 600 BCE.

There are a number of other optical devices that use refracting lenses. One is the refracting telescope, figure 4.40. It may be compared to the reflecting telescope of Isaac Newton, figure 4.8. Galileo (1564–1642) was the first person to use the refracting telescope for astronomical purposes. He discovered the moons of Jupiter, the phases of Venus, the mountains on the Moon, sunspots and thousands of stars

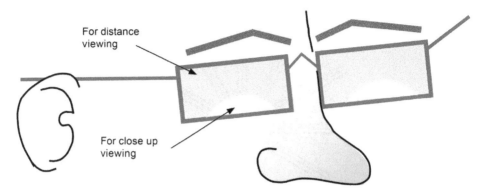

Figure 4.38. Bifocals, invented by Benjamin Franklin, correct for distant and nearby viewing inadequacies. According to my eye doctor, everybody needs bifocals sooner or later.

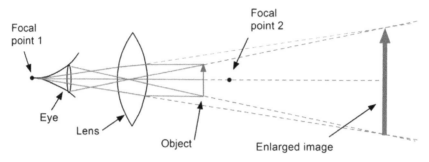

Figure 4.39. A magnifying glass is a double convex refracting lens with the object to be magnified placed between the lens and one focal point. When a viewer looks into the lens from its other side, they see an enlarged, non-inverted image.

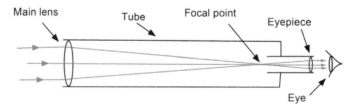

Figure 4.40. The refracting telescope. An enlarged, inverted image is observed.

which are too dim for the unaided eye to see. These discoveries provided incontrovertible proof of the validity of the heliocentric theory of the solar system.

Telescopes provide three services to the observer: they gather more light than the unaided eye, they provide magnification and they improve resolution, which is the ability to visually distinguish two objects which are close together. On the other

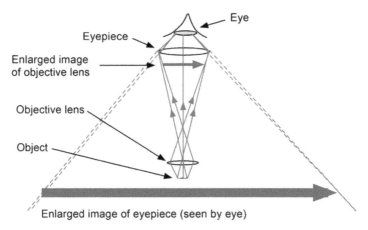

Figure 4.41. A microscope consists of an objective lens and an eyepiece. The object is placed just beyond the focal length of the objective lens; its image is thus magnified by the eyepiece, resulting in a very large, inverted image.

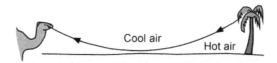

Figure 4.42. The camel looks down and sees the top of the palm tree. The light from the tree refracts upwards due to the temperature differences in the air. A mirage results.

hand, the *microscope* (Galileo constructed one in 1625) mainly magnifies very small objects, figure 4.41.

Cameras and binoculars, figure 4.28, are two additional optical instruments that use refracting lenses. Earth-observing satellites have excellent cameras on board. These are capable of making a readable image of a newspaper on the ground 100 miles below.

4.3.3 Mirages

Other interesting effects are due to refraction. One of these is mirages, figure 4.42. The light ray from the palm tree bends by refraction as it travels through the different layers of air. The hot air directly over the sand has a lower density than the cooler air over it; the speed of light is different in the two regions.

A similar effect causes 'water' to appear on asphalt roads on sunny summer days. The 'water' is actually a mirage; it is the image of the bottoms of nearby clouds. At night the air at ground level is cooler than the air just above it. Sometimes UFO 'observations' are caused in this way. Submarines take advantage of layers of water of different temperature to escape sonar detection by enemy destroyers, figure 4.43. Refraction of the sonar pulses makes the submarine 'invisible'. Sounds are some-times clearly heard from across a lake in the evening as the sound waves bounce off

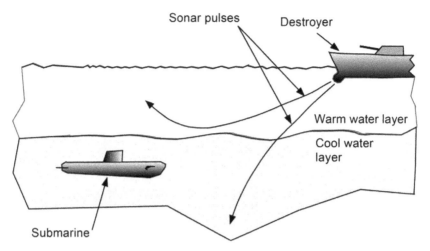

Figure 4.43. Submarines escape detection by cruising in the cool layer just below a warm layer. The sonar pulses are refracted upward at shallow angles of incidence and downward at higher angles, resulting in a 'shadow zone'.

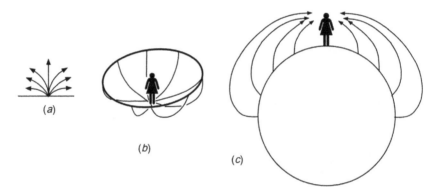

Figure 4.44. Supercritical refraction as postulated to occur on Venus. (a) Light rays from an object bend up due to refraction. (b) An observer sees objects about them. (c) To the observer, it appears as if they are in a bowl-shaped depression.

the water and then refract back down from the layers of warmer and cooler air just over the surface of the lake.

4.3.4 Supercritical refraction

Supercritical refraction is a condition that may occur on Venus, whose atmosphere is hot (over 863°F, 462°C) and dense (the surface pressure is 100 times that on Earth). Because the density of the atmosphere diminishes rapidly as one rises above the planet's surface, refraction effects are predicted to be extreme. As figure 4.44 shows, an observer on Venus would see a bowl-like effect, with the horizon tilted upward. As the observer moved around on the surface of Venus, they would always seem to be at the bottom of a large bowl-like depression. One prediction states that the effect is so large that as the observer looks forward, they will see the back of their own head! Two Soviet spacecrafts have sent pictures back to Earth from the surface of Venus and supercritical refraction is not apparent in the pictures. This may be due, however, to the closeness of the cameras to the surface. The postulated existence of supercritical refraction on Venus is still open to question.

4.3.5 Dispersion

Dispersion is defined as the dependence of the degree of refraction on wavelength. Dispersion is directly responsible for rainbows and other colorful phenomena. Consider a red and a blue beam of light falling on a piece of glass (figure 4.45(a)).

Dispersion causes the blue beam to refract more than the red beam, figure 4.45(a). Now if we shine a beam of white light on the glass, a spectrum of colors appears ranging from red to violet, figure 4.45(b). The colors vary smoothly from the long wavelength red to the short wavelength violet. We have separated the colors by convention into red, orange, yellow, green, blue, indigo and violet. Prisms conveniently demonstrate dispersion; refraction takes place twice and a greater spread of the colors appears, figures 4.46 and 3.6.

What the human eye perceives as 'white' is an equal mixture of all the visible wavelengths, figure 4.47(a). For only one wavelength, one color is seen, figure 4.47(b). 'Pink' is perceived if all of the colors are present somewhat equally, with a little extra red component, figure 4.47(c).

The optical *spectroscope* breaks up a beam of light into its individual components for observation, figure 4.48. The source of light for the optical spectroscope can be in the laboratory, or it can be a distant star or galaxy. Astronomers make great use of

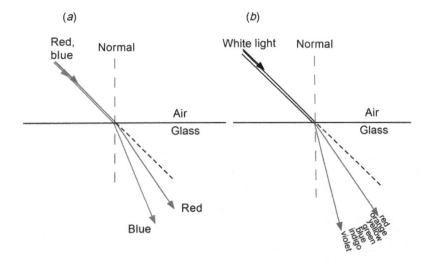

Figure 4.45. Dispersion. (a) A red beam and a blue beam are refracted differently. (b) White light is broken up into the spectrum: red, orange, yellow, green, blue, indigo and violet.

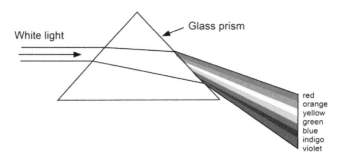

Figure 4.46. The prism uses refraction to disperse white light into its colors.

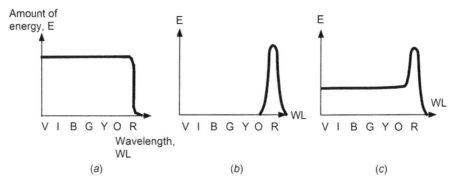

Figure 4.47. Spectral plots of (a) white, (b) red and (c) pink.

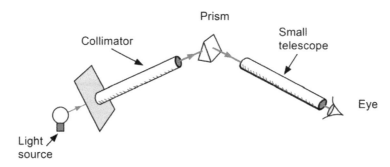

Figure 4.48. An optical spectroscope.

spectroscopy. A star's spectrum tells us its atomic makeup, temperature, age, size, mass, distance, if it has a companion star or planet, and its velocity of approach or recession with respect to us. Our Sun's spectrum revealed the presence of helium 80 years before any helium was found on Earth.

The rainbow is a natural phenomenon which beautifully demonstrates dispersion.

The primary and secondary bows are formed as shown in figure 4.49: sunlight enters the raindrops and is refracted twice for the primary bow and three times for the secondary bow. Note that the color scheme is reversed in the secondary bow as opposed to the primary (figure 4.50). Because of the geometry of spheres

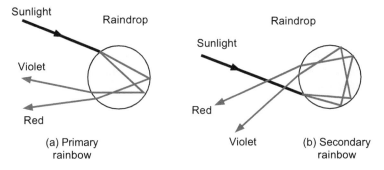

Figure 4.49. Primary (a) and secondary (b) rainbows.

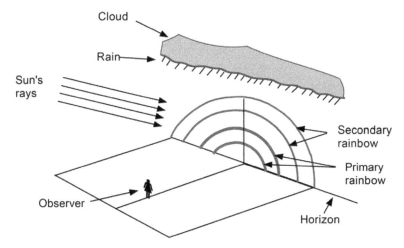

Figure 4.50. How the primary and secondary rainbows appear to an observer. Note the reversal of colors in the secondary bow.

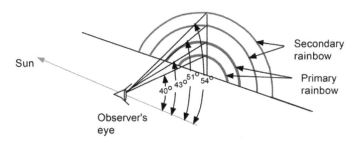

Figure 4.51. Rainbow angles.

and the optical path lengths through them, cones of light exit the drops in certain directions (figure 4.51). Similar colorful atmospheric effects—lunar and solar halos and sun dogs—are due to refraction and reflection from tiny atmospheric ice crystals.

4.4 Diffraction and interference

Diffraction is the bending of waves around an obstacle, figure 4.52. It is more apparent in the case of sound waves than light waves because the region where diffraction can be observed is about the same size as one wavelength. Sound waves are much longer than light waves, so diffraction is more obvious for sound rather than light.

Interference occurs because of *superposition*: the adding of two waves (figure 4.53).

There are two types of interference: constructive and destructive (figure 4.54).

Here follow some examples of diffraction and interference.

Auditorium dead spots. Figure 4.55 shows an auditorium with a dead spot. When the singer A produces a sound, one wave ABC hits the wall at B and arrives at C 'out

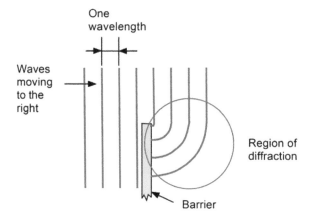

Figure 4.52. Diffraction occurs when the wave (water or light) strikes a barrier and spreads out.

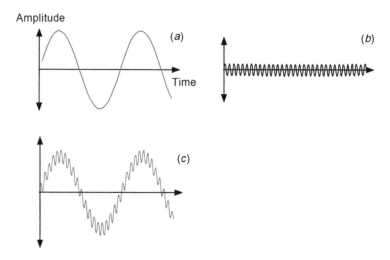

Figure 4.53. Superposition: waves (a) and (b) are added to obtain a combined wave (c).

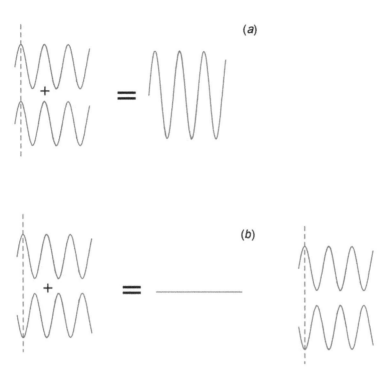

Figure 4.54. Interference: (a) constructive and (b) destructive.

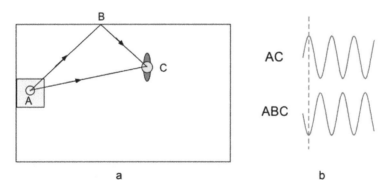

Figure 4.55. (a) An auditorium with a 'dead spot'. (b) A sound wave traveling along the path ABC will exactly cancel the sound wave traveling along path AC.

of phase' with the wave AC that goes directly to C, the listener. Destructive interference occurs and C cannot hear the sound. Dead spots can be minimized by careful design and by choosing sound absorbing materials for walls and ceilings.

The two slit experiment validated Thomas Young's theory that light is a wave. When coherent light is shone on two slits, we observe an interference pattern, figure 4.56. The same experiment using particles such as electrons can also yield interference patterns, a quantum physics phenomenon.

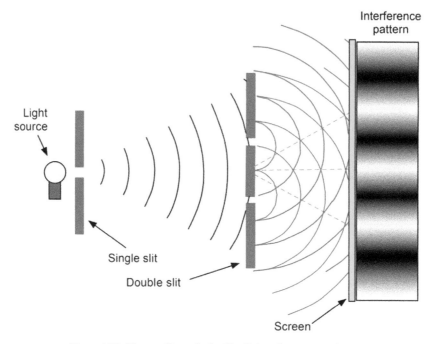

Figure 4.56. Thomas Young's double slit interference experiment.

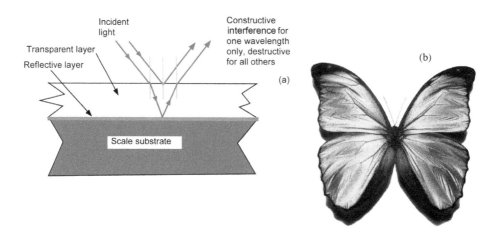

Figure 4.57. (a) Iridescence produced by constructive and destructive interference. (b) The Blue Morpho butterfly. https://commons.wikimedia.org/wiki/File:Blue_morpho_butterfly.jpg. This image has been obtained by the author from the Wikimedia website where it was made available under a CC BY-SA 3.0 licence. It is included within this article on that basis. It is attributed to Gregory Phillips.

Iridescent colors. Displayed by some animals, iridescent colors are produced by interference (figure 4.57). In butterflies, snakes and lizards and hummingbirds, the scale thickness is such that constructive interference occurs for one color only. The resulting colors are much more brilliant than those produced by dyes and other pigments.

Coated lenses. Coated lenses maximize the light entering an optical device such as binoculars or cameras. A coating on the primary lens's outer surface causes destructive interference of the reflected light. Almost 100% of the incoming beam enters the lens. For this to happen, the coating must be one quarter of a wavelength thick, figure 4.58. This works best for only one wavelength. The bluish tint of coated lenses suggests that some other color is the choice for optimum absorbance. This is in fact the case—yellow light is the preferred color. The coating layer is rather soft and easily scratched. So never clean your coated lenses with anything other than a soft camel's hair brush.

Holograms are produced by the interference of a reference beam and a beam that strikes the object, figure 4.59. The resulting interference pattern is recorded on a film. Shining a laser on the film gives us a three-dimensional image, figure 4.60.

Diffraction gratings. Bouncing light off of a diffraction grating produces constructive interference only in certain well-defined directions, figure 4.61. Thus, when you look at a diffraction grating, you will see certain colors at one spot and others at other spots.

Soap bubbles and oil films. Soap bubbles and oil films also produce their colors using interference. Figure 4.62 shows how varying thickness of the bubble and an oil film can produce different colors.

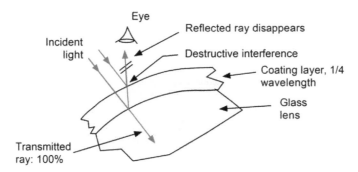

Figure 4.58. The coated lens: a practical use of interference.

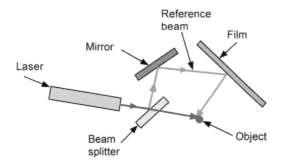

Figure 4.59. Production of a hologram. An interference pattern is formed on the film.

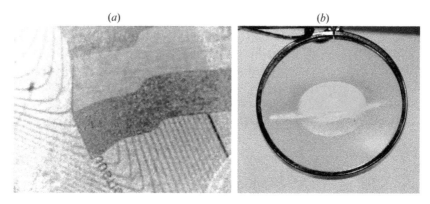

Figure 4.60. Holograms. (a) A close-up photograph of a hologram's surface. (b) The planet Saturn.

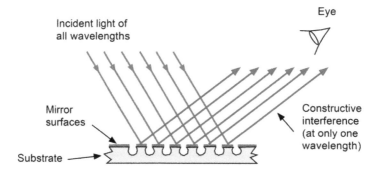

Figure 4.61. A diffraction grating produces brilliant colors.

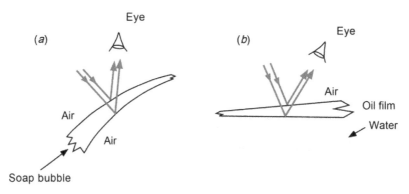

Figure 4.62. Interference producing colors in (a) soap bubbles and (b) oil films on water.

Scattering is a combination of interference and diffraction, figure 4.63. Scattering is most efficient when the wavelength of the scattered wave is about the same size as the particle doing the scattering. The Earth's sky is blue and sunsets are red because of scattering, figure 4.64.

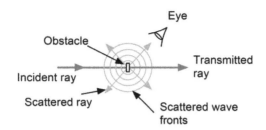

Figure 4.63. The scattering of light from an obstacle.

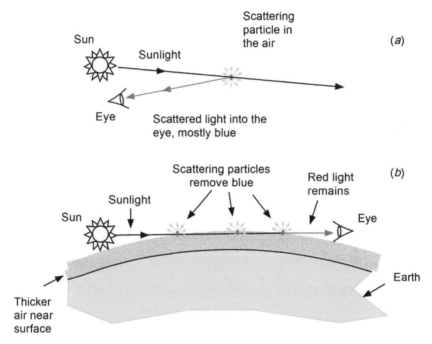

Figure 4.64. Why the sky is blue (a) and sunsets are red (b): scattering is responsible.

The water droplets in clouds are much larger than air molecules. So light scattering from clouds is non-selective, and all wavelengths are scattered equally. Thus clouds appear white.

Mars' sky is not blue but pink. In this case, absorption in Mars' atmosphere by reddish dust particles causes the pink sky coloration. Since Mars often has violent windstorms and a red soil, and since it hardly ever rains on Mars, the sky is full of tiny red dust particles.

Polarization occurs only for transverse waves. All of the other wave phenomena that we have discussed occur for all types of waves. Figure 4.65 shows an unpolarized wave hitting a polarizer and producing a polarized wave. When this polarized wave hits another polarizer aligned in the same direction, the polarized wave passes right through the polarizer. However, if the polarized wave hits a polarizer aligned perpendicularly to

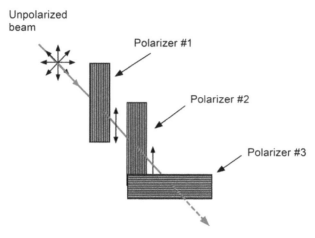

Figure 4.65. Polarizing a light beam.

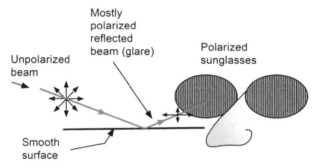

Figure 4.66. Vertically polarized sunglasses block the (mostly) horizontally polarized glare.

it, the polarized wave will be stopped. E H Land invented a clever way of producing polarizing material using plastic. This material is used widely in sunglasses to cut down on glare. Polarized sunglasses can reduce glare because when light glances off a smooth surface it is partially polarized horizontally, as shown in figure 4.66.

Polarization is used also in the liquid crystal displays (LCDs) of digital watches, TVs and calculators. You can check this with your polarized sunglasses: rotate the watch or calculator and watch its display turn black!

4.5 The Electron microscope and quantum physics

An important application of interference is the electron microscope. In 1897, British physicist J. J. Thompson discovered the electron, one of the building blocks of atoms. Since the electron is a quantum particle, we can use electrons to make a microscope of unparalleled power. In 1905, Einstein explained an unusual physical phenomenon called the 'photoelectric effect.' It turns out that when light is shined on a piece of metal, electrons are ejected from the metal's surface. Only light of wavelengths shorter than a certain value (which depends on the type of metal being used) was able to eject the electrons. Even a very weak beam of light of the correct wavelengths would eject a few electrons. However, if the wavelengths were too long—no electrons would leave the metal, even if the intensity of the beam was very powerful.

This was a BIG puzzle for the physicists of the day. Albert Einstein solved the paradox in 1905. He postulated that the light impinging on a metal surface should be thought of not as a wave but as a beam of individual particles called photons. The energy of each photon was quantized and was inversely proportional to its wavelength. Shorter wavelengths meant more powerful photons. Einstein then went on to produce a formula which correctly explained the experimental results. For this work, Einstein was awarded the Nobel Prize for Physics in 1915.

Quantum physics presents us with phenomena that are hard to fathom, but nevertheless are backed up with experimental results. One of my favorites is the double slit experiment, figure 4.56, page 4-29. Now imagine that the light source is substituted with a source of electrons. An interference pattern is still observed. The inescapable conclusion is that the electrons are, as Einstein said, exhibiting their wave-like properties and interfering with each other as if they were waves.

Where it gets spooky, is that it is experimentally possible to slow the electron beam down so much that the next electron will not leave the electron source until after the previous electron has already hit the screen. An interference pattern is still observed! In other words, each electron is interfering *with itself*, that is, each electron is going through both slits at the same time!

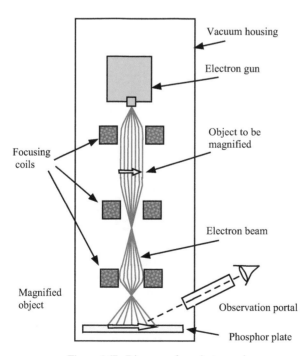

Figure 4.67. Diagram of an electron microscope.

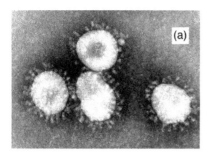
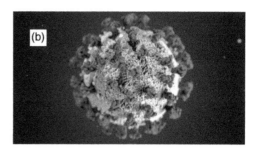

Figure 4.68. (a) An electron micrograph of the corona virus. Photo: CDC/Dr Fred Murphy; Sylvia Whitfield. (b) This CDC illustration shows the surface of a corona virus, identified as 2 (SARS-CoV-2); it is the cause of the coronavirus disease 2019 (COVID-19) in Wuhan, China in 2019. Photo credit: Alissa Eckert, MSMI; Dan Higgins, MAMS.

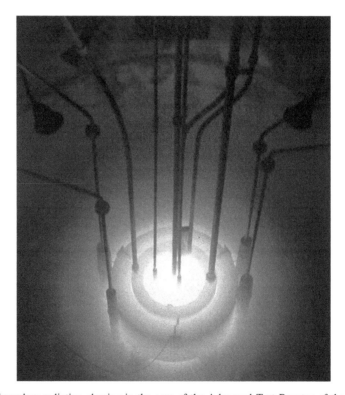

Figure 4.69. Cherenkov radiation glowing in the core of the Advanced Test Reactor of the Reed Research Reactor, Reed College, Oregon. This image has been obtained by the author from the Wikimedia website https://commons.wikimedia.org/wiki/File:Cerenkov_Effect.jpg, where it is stated to have been released into the public domain. It is included within this article on that basis.

The electron microscope depends on the electrons' wave-like properties, figure 4.67. The electron beam can be concentrated and focused in a manner similar to beams of light.

The electron microscope is an essential instrument for medical research; an example is given in figure 4.68.

4.6 Cherenkov radiation: faster than light

While it is true that nothing can travel faster than light in a vacuum, electrons can indeed travel faster than light in water. Pavel Cherenkov discovered that light is emitted when an electron in water has a phase velocity greater than that of light in water. The phase velocity is equal to the speed of radiation (light) times the index of refraction (section 4.3). In pool-type nuclear reactors, high-energy electrons are released into the water by the nuclear fission reaction. These super-fast electrons travel at high speeds, even faster than light can travel *in water*. The shock wave that the electrons generate is like the sonic boom of a supersonic airplane going faster than the speed of sound. Cherenkov radiation is somewhat like the sonic boom. Figure 4.69 shows the eerie blue glow of Cherenkov radiation emitted in the core of the Advanced Test Reactor at the Idaho National Laboratory in Arco, Idaho. Cherenkov shared the Nobel Prize for Physics in 1958.

This is the end of the chapter on the properties of waves and their effects on us. Now we can proceed to our next chapter: the phenomenon we know as hearing.

Chapter 5

Hearing: the perception of sound

In this chapter, we will look more closely at the perception of sound in general and musical sounds in particular. By learning about some of the physics behind the sounds of music, we will increase our enjoyment and appreciation of these sounds.

5.1 Sound intensity

The ear can hear extremely soft sounds and extremely loud sounds. When we hear a sound, we are detecting differing levels of air pressure. We measure sound intensity in units of watts per square meter ($W\ m^{-2}$). The normal human ear can hear sounds as low as $1/1\ 000\ 000\ 000\ 000\ W\ m^{-2} = 10^{-12}\ W\ m^{-2}$ and as high as $1\ W\ m^{-2}$ (sounds louder than this cause physical pain and permanent damage). This is a tremendous range, a trillion to one, and it is truly astounding. This wide range is also very inconvenient to work with because the numbers vary so much, even with scientific notations. As a result, scientists developed a new unit to make things easier: the *decibel* (dB). Using the decibel, the range of hearing goes from 0 dB at the bottom of the range up to 120 dB at the threshold of pain. This scale is 'logarithmic' and is a little different from the regular numbering system that we use in our everyday lives. Figure 5.1 lists the intensities and loudness for some common sounds.

The definition of the decibel is given as:

$$L = 10 \log(I/I_0)\ \text{dB}$$

where L is the loudness in decibels, log is the logarithm function, I is the sound intensity in $W\ m^{-2}$ and I_0 is the reference sound intensity, $10^{-12}\ W\ m^{-2}$.

If two sounds differ by 10 dB, their loudness difference is 10 times. In other words, if one singer sings with a loudness of, say, 70 dB, ten similar singers would have a loudness of 80 dB. 100 singers would produce a loudness of 90 dB, 1000 singers would produce a loudness of 100 dB, and so on. Only two singers would have a loudness of 73 dB; an increase of 3 dB represents a doubling of the loudness, figure 5.2.

doi:10.1088/978-0-7503-3207-1ch5

Intensity (w/m^2)	Loudness (dB)	Description	Music designation
100	140	Airplne taking off	
10	130	Pain	
1	120		
10^{-1}	110		
10^{-2}	100	Roller coaster	f f f
10^{-3}	90		
10^{-4}	80	Heavy auto traffic	f forte
10^{-5}	70	Conversation	
10^{-6}	60		p piano
10^{-7}	50	Quiet room	
10^{-8}	40		p p p
10^{-9}	30	Whisper	
10^{-10}	20		
10^{-11}	10		
10^{-12}	0	Reference	

Figure 5.1. The intensity-loudness chart.

1 Singer, 70 dB   2 Singers, 73 dB   10 Singers, 80 dB

Figure 5.2. Singer loudness.

Our perception of loudness has some interesting aspects. For example, the *just noticeable difference* of two sounds depends on the frequency and loudness of the sound, figure 5.3. Also, our hearing sensitivity depends on frequency and is slightly different for each of our two ears (diplacusis).

Masking is an example of an auditory illusion. If a perfectly audible sound has a louder sound played at the same time, the two sounds will be distinguishable only if they are widely separated in frequency. As the frequency of the louder sound approaches that of the softer sound, the softer sound disappears and is no longer heard. This is very important for music because it has a large effect on orchestration. For example, when the brasses are playing fortissimo, the bassoonist can't even hear what he/she is playing.

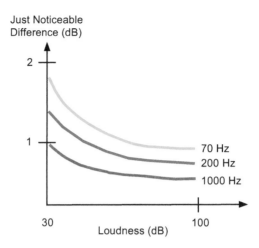

Figure 5.3. The just noticeable difference between two sounds depends on their frequencies and loudness.

When a single frequency sound (*pure note*) comes into the ear, it travels down the auditory pathway and eventually enters the cochlea. The vibrations then excite only one area along the cochlea of the basal membrane. The magnitude of the vibrations and their location on the basilar membrane tell the brain what the frequency and intensity of the sound is. If another pure note is sounded and enters the ear, the same thing will happen, but a different location on the basilar membrane will be excited. If the two notes are similar enough in frequency, the two vibrating areas of the basilar membrane will overlap and only the louder of the two sounds will register in the brain.

We measure hearing ability with an *audiometer*. Hearing loss can be temporary, as when a firecracker goes off near you, or it can be induced by constant exposure to loud sounds. Figure 5.4(a) shows the hearing range of a normal ear and figure 5.4(b) shows the hearing ranges of two common situations. First, the loss of hearing ability in a specific range for a pistol shooter is shown. Second, the loss of the ability to hear the higher frequencies as a person ages, called *presbycusis*. The word presbycusis comes from the same root 'presby' as presbyterian, a protestant religious denomination based on governance by elders.

Hearing impairment may be due to age, birth defects, ear infections or injury. Defective ossicles can be replaced with plastic devices and artificial cochlea implants are a reality (see paragraph 2.4.3). Engineers are constantly improving electronic hearing aids.

Tinnitus, a ringing in the ears, can be a very serious condition. Some sufferers have even resorted to suicide for the ultimate permanent relief. In at least one case, a patient demanded, and received, a surgical procedure to sever his auditory nerve. Unfortunately for him, the operation did not stop the tinnitus although his hearing was completely gone. The origin of the tinnitus was further along the auditory pathway than his auditory nerve, originating somewhere in his brain.

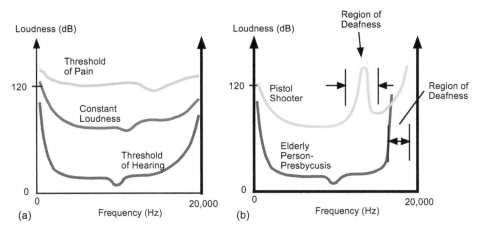

Figure 5.4. (a) The hearing ability of a normal person. The dip in the threshold of hearing reflects a sensitivity to a frequency similar to a baby's cries and to the sirens of emergency vehicles. (b) The regions of deafness typical of pistol shooters and senior citizens.

5.2 Tones

A tone is a kind of sound that is loud enough, has a constant enough frequency and lasts long enough (duration) to have three certain characteristics: *loudness, quality* and *pitch*. We have already seen that we measure loudness in decibels. Quality or timbre permits a tone to be distinguished from other tones of the same fundamental frequency and loudness. A *pure tone* or simple tone looks like a sine wave when displayed on the oscilloscope, for example. All other tones are complex; they look more or less bumpy, figure 5.5. Any complex tone can produced by adding a number of pure tones. We call these pure tones '*partials*'. *Synthesis* is the process of adding up pure tones of the proper amplitude and frequency to make a complex tone. *Analysis* is the separating of a complex tone into its constituent partial tones. The frequency and amplitude of the partials of a particular complex tone determine the quality of that tone. A complex tone, if it is steady for long enough, can have a fundamental frequency, figure 5.5(c). If a tone is steady, or stable, then we call its partials harmonics—the ratio of the frequencies of the partials and the fundamental frequency is the ratio of integers: 1:2, 2:3, 4:5, 3:4, etc.

In examining the structure of a complex tone, it is helpful to look at the spectrum of the tone. The spectrum is a listing or picture of the frequencies and amplitudes of the partials. Figure 5.6 shows the spectra of a pure tone, a sawtooth wave, a square wave, a triangular wave and a repeating pulse. Figure 5.7 lists these in tabular form. Although each type of wave has the same fundamental frequency in this example, each has an entirely different sound.

A complex tone (a sawtooth wave in this case) can be synthesized or produced by adding its harmonics, figure 5.8.

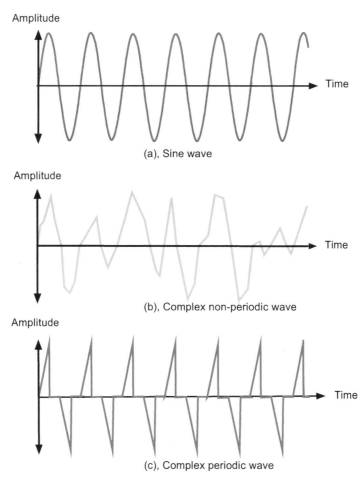

Figure 5.5. Amplitude versus frequency for three sounds: (a) a pure tone—a sine wave produced by a tuning fork, for example; (b) a complex, non-periodic sound—crowd noise, for example; and (c) a complex periodic tone—a trumpet playing a single note, for example.

5.2.1 Frequency and pitch

At first glance, frequency and pitch might appear to be the same—but they are not. We may define the frequency of a sound as its number of cycles per second and its pitch as the perceived frequency. The audible frequency spectrum covers 15–15 000 Hz; pitch discrimination is poor in both the lower and upper portions of this range. Presbycusis is the progressive inability to hear the highest frequencies as we age. The frequency span of musical sounds goes up to only 10 000 Hz. So presbycusis is not very important to musicians. The highest note on a piano is 4186 Hz; because of harmonics, we need to hear a little higher in frequency than this, but not much.

Three factors affect pitch: *diplacusis* (the fact that our two ears hear slightly differently), *duration* and *intensity*. We have already discussed diplacusis. Duration also affects our ability to distinguish pitch; over a long period of time the pitch of a

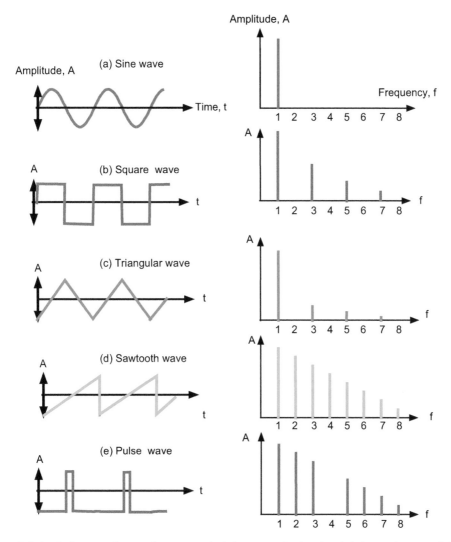

Figure 5.6. Analyzing several types of waves reveals their spectra, that is, what their harmonics are and their harmonics' amplitudes and frequencies. (a) A pure tone, that is, a sine wave, (b) a square wave, (c) a triangular wave, (d) a sawtooth wave and (e) a pulse wave.

constant-frequency sound seems to waver, or wander around. If a constant-frequency sound is being listened to, and the volume of the sound is suddenly increased, the pitch of the sound will appear to shift downward at low frequencies and will appear to shift upward at high frequencies. Absolute pitch is the uncanny ability of some people to be able to associate any sound that they hear with a tone on the musical scale.

	Amplitude of harmonic	Frequency of harmonic [a]
Sine Wave (pure tone)	1	1
Square Wave	1	1
	1/3	3
	1/5	5
	1/7, etc.	7, etc.
Triangular Wave	1	1
	1/9	3
	1/25	5
	1/49	7, etc.
Sawtooth Wave	1	1
	1/2	2
	1/3	3
	1/4	4
	1/5, etc.	5, etc.
[a]times fundamental frequency		

Figure 5.7. A tabular listing of the amplitude and frequencies of the harmonics of four tones: a pure tone (a sine wave); a square wave; a triangular wave; a sawtooth wave; and a sawtooth wave (top to bottom).

5.2.2 Musical scales

The musical scale is a certain number of designated frequencies or tones. A note is an individual frequency. The wave of a pure tone is a sine wave. *Consonance* is a pleasing musical sound made by combining two or more notes. The Greeks studied music theory with a device called a *monochord*, figure 5.9. The fundamental wavelength of a vibrating string is proportional to the length of the string. The frequency of the string's sound is inversely proportional to the wavelength of the sound.

A long string will have a lower frequency than a short string. The Greeks found that when the length ratio of the two segments, ℓ_1/ℓ_2, is the ratio of two integers, then a pleasing sound (consonance) is produced by vibrating the strings. Some of the ratios have names:

1:1 unison
2:1 or 1:2 octave
3:2 or 2:3 fifth
4:3 or 3:4 fourth.

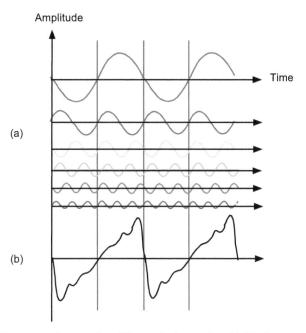

Figure 5.8. Synthesizing a complex tone by adding up its harmonics. (a) The harmonics. (b) The resulting complex tone, which will be a sawtooth wave when all of the harmonics are added.

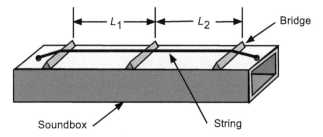

Figure 5.9. The monochord.

The piano is cleverly arranged to optimize consonance, figure 5.10. There are eight white keys between each octave and five black keys for a total of twelve notes between the octaves. The highest note of each octave has twice the frequency of the lowest note. If we knew the frequency interval between each note and the frequency of just one note, we could construct the entire scale. The *even tempered scale* (which is the one we use) gives us these two things. The frequency interval between two notes is called the *tempered semitone* and is designated by the letter '**a**'. The reference note for the even tempered scale is designated as the note A above middle C, this note has a defined frequency of 440.0 Hz. Since we get to frequency f_2 from frequency f_1:

$$f_2 = \mathbf{a}\, f_1.$$

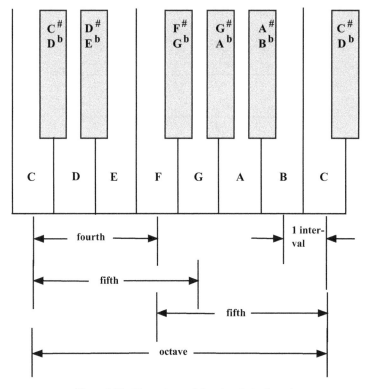

Figure 5.10. One octave of the piano's keyboard.

If we start at the lowest note of an octave and proceed according to

$$f_3 = \mathbf{a}\, f_2 = \mathbf{a}\,\mathbf{a}\, f_1 = \mathbf{a}^2 f_2$$

and so on, then at the high end of the octave

$$f_{12} = \mathbf{a}^{12} f_1 = 2 f_1$$

and

$$\mathbf{a}^{12} = 2.$$

Using our trusty calculator $\mathbf{a} = 2^{(1/12)} = 1.059\ 4631$.

Between two adjacent notes, the number of cycles per second is not a constant number. However, when the frequencies of two adjacent notes are divided, the number $\mathbf{a}$ always results. The note middle C is nine notes below A. So middle C has a frequency

$$f = \mathbf{a}^{-9} \times (440.0) = (0.594\ 603\ 58) \times (440.0) = 261.625\ 57\ \text{Hz},$$

or 261.6 Hz rounded off. The frequency of C#, for example, is

$$f = \mathbf{a} \times 261.6\ \text{Hz} = 271.2\ \text{Hz}.$$

C	C#	D	D#	E	F	F#	G	G#	A	A#	B	C
1	a	a^2	a^3	a^4	a^5	a^6	a^7	a^8	a^9	a^{10}	a^{11}	a^{12}
1.000		1.122		1.260		1.414		1.587		1.782		2.000
	1.059		1.189		1.335		1.498		1.682		1.888	

Figure 5.11. One octave's semitones.

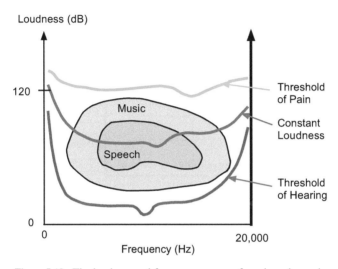

Figure 5.12. The loudness and frequency ranges of music and speech.

Figure 5.11 shows the semitones of one octave.

Other musical scales have been set up throughout history. We use the even tempered scale because it sounds better to us than the other scales. By deviating from the even tempered scale in a carefully controlled way, we can accomplish *intonation*. *Chords* are a series of specific notes sounded simultaneously for a pleasing effect. *Harmony* is a similar musical technique.

5.3 Musical instruments

Music adds much to the enjoyment of our lives. In this section, we will study the physics of how musical notes are produced. We will leave it to the musicians to actually use this knowledge to make music.

Music and speech lie well within the volume and frequency ranges of our hearing, figure 5.12. All musical instruments have two things in common: they each have a way of producing vibrations and they each have a way of coupling the vibrations to the surrounding air. This produces sound waves that we can hear. Figure 5.13 shows the frequency ranges for various musical instruments.

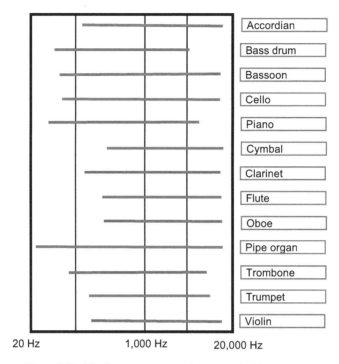

20 Hz   1,000 Hz   20,000 Hz

Figure 5.13. The frequency ranges of some musical instruments.

5.3.1 Vibrating strings

We can separate musical instruments into groups according to the source of vibrations in each type. Perhaps the simplest source of vibrations is the vibrating string. A vibrating string produces a standing wave with a number of different modes of vibrations, figure 5.14. Note that the modes have frequencies that are multiples of the fundamental frequency. When we add these modes together, as in a plucked guitar string, the resulting combination is a stable complex tone. The partials are harmonics of the fundamental frequency. The string's vibration frequency depends on the tension of the string and inversely on the mass per unit length of the string and the length of the string. We tune the instrument by adjusting these physical properties of its strings.

We divide the vibrating stringed instruments into three basic types: (1) *plucked* or strummed strings (guitar, harp, harpsichord), figure 5.15; (2) the *bowed strings* (violin, etc), figure 5.16; and (3) the *struck strings* (piano, dulcimer, clavichord), figure 5.17. For an acoustic guitar, figure 5.15, all six of the strings have the same length, but their diameters and materials of manufacture are different. This enables the guitar to have six individual notes. The frets allow the guitarist to change the effective length of each string, and thus the frequency of each string. The body and sound hole of the guitar couple its sounds to the surrounding air. For an electric guitar, electromagnetic pickups detect the strings' vibrations. An amplifier and speakers complete the instrument.

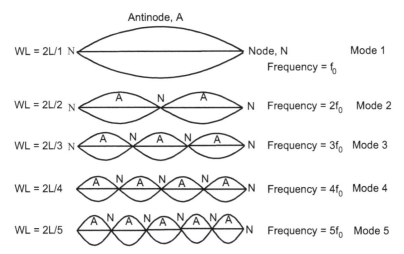

Figure 5.14. The vibrational modes of a stretched string.

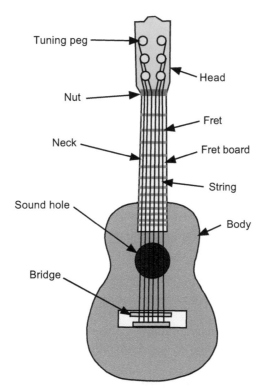

Figure 5.15. The parts of an acoustic guitar.

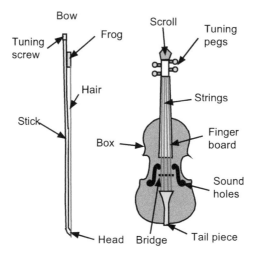

Figure 5.16. Parts of the violin.

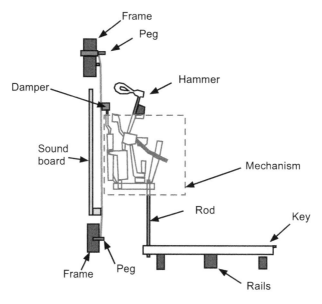

Figure 5.17. The piano key assembly, for an upright piano.

In the bowed instruments, the action of the bow is one of stick-slip-stick-slip across the strings. This is similar to scraping one's fingernails across a chalkboard to produce a screeching sound.

In the bowed string instruments, a wooden box couples the complex vibrations of the strings to the surrounding air. Sound waves that we can clearly hear are the result. The musicians rub rosin (hardened pine sap) on the bow fibers. This greatly enhances the stick-slip-stick process by increasing the friction between the bow and the strings. Note that the lack of frets means that the musician must use finger

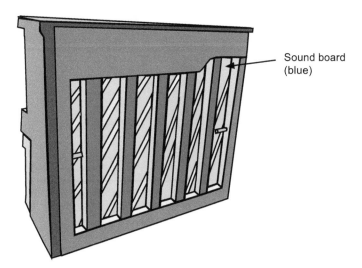

Sound board
(blue)

Figure 5.18. A back view of an upright piano shows the sound board, a thin piece of wood. Grand pianos must also have a sound board.

pressure on the strings and finger board to change the effective length of the strings. In this way, various tones can be produced. Very complex tones can be generated by bowing more than one string at a time. Unlike the plucked stringed instruments, the bowed stringed instruments can produce sustained sounds of varying loudness. The bowed strings can also be plucked; this is called the *pizzicato* technique.

We call the third class of stringed instruments the *struck string* instruments, typified by the piano. Small felt hammers strike the strings, figure 5.17. The complicated key–hammer–string mechanism enables the pianist to produce a soft sound with a soft touch of the key, a loud sound with a strong hit on the key, and a sustained tone by holding the key down. With its versatility, ability to produce sounds of varying volume and duration and wide frequency range (seven octaves), it's no wonder that the piano has become one of the most popular of musical instruments. In 1709, Bartolommeo Cristofori invented the piano in Florence, Italy. It was originally called the fortepiano ('forte' means loud and 'piano' means soft in Italian). When one of the 88 keys is struck, the hammer hits the strings for that note. The vibrations are coupled to the air of the room by the sound board, figure 5.18.

A steel frame is necessary to counterbalance the great tension of the strings. The invention of the piano, a variation of the existing harpsichord, had to wait on the progress of metallurgy to make steel of sufficient strength. The steel frame, which looks a lot like a harp, is mostly responsible for the heavy weight of pianos. To obtain the requisite volume for the higher notes and their shorter strings, there are three strings for each key instead of the two strings for each of the lower notes. To obtain the required mass per unit length for the lowest notes, each string is wrapped with copper wire.

The black and white arrangement of the piano keys is widely used by other, mostly electronic, instruments, such as the piano tie, figure 5.19. This can be worn as a necktie, runs on batteries, has a small speaker in the knot and can be purchased on-line.

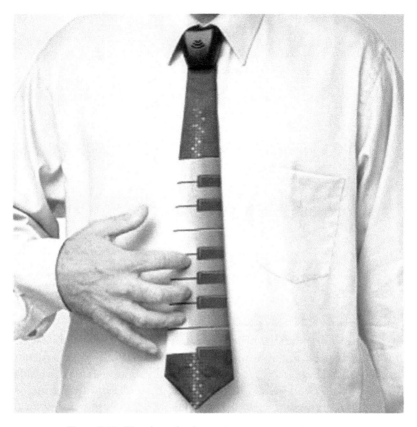

Figure 5.19. The piano tie plays notes over a one octave range.

5.3.2 Reed instruments

Strings stretched between two points can be set to vibrate and so can columns of air. The columns can be open at one end, or closed at both ends, figure 5.20. The column closed at both ends and the column open at both ends have the same type of modes of vibration as the vibrating string (all harmonics present). But, the column open at one end has the even modes missing. The longer a pipe is, the lower its frequency will be. A look at a large pipe organ shows how important pipe length is for producing different notes.

We divide vibrating air column instruments into four basic types: (1) the *air reeds*, (2) the *mechanical reeds*, (3) the *lip reeds* and (4) the *vocal cord reeds* (the voice). In the air reeds, the 'reed' refers to the methods which can cause air to vibrate in a column. Originally 'reed' referred to a mechanical device such as is found in harmonicas; the term has been broadened to include the air at the top of a bottle, figure 5.21(a). Other air reed instruments include the recorder, flute and whistle.

The *mechanical reeds* have either a *single reed*, figure 5.21(b) and (c), or *double reed*, figure 5.21(d). The harmonica, accordion, clarinet, saxophone and bagpipe are single reed instruments whereas the oboe, English horn and bassoon are double reed instruments.

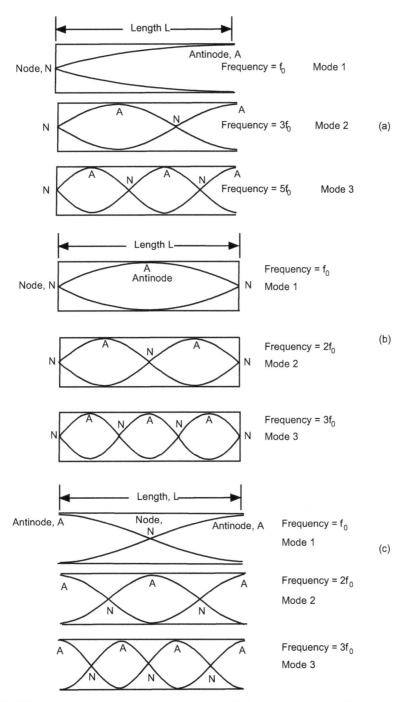

Figure 5.20. Vibrational modes of pipes (a) open at one end, (b) closed at both ends and (c) open at both ends.

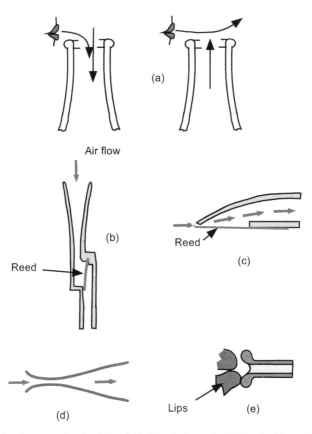

Figure 5.21. (a) The air reed, a bottle. (b) and (c) The single reed. (d) The double reed. (e) The lip reed.

Lip reed instruments, figure 5.20(e), use pursed lips to set up vibrations in bugles, trumpets and cornets. Valves change the lengths of the air columns, producing different notes.

In the case of the *human voice*, the vocal cords act as reeds and set up vibrations in the pharyngeal cavity, figure 5.22. The lips, teeth, tongue and palate control the sounds and form the words of speech. The famous singer Mariah Carey has an incredible vocal range of over five octaves: E2 to G#7, figure 5.23.

5.3.3 Percussion instruments

Percussion instruments are those which are struck to produce sound. There are two main categories: the *definite pitch* instruments (bells, tuning forks and glockenspiels) and the *indefinite pitch* instruments (snare drums and cymbals). Figure 5.24 shows the Liberty Bell, a definite pitch percussion instrument located in Philadelphia, Pennsylvania. Unfortunately, the bell cracked the first time it was rung. It was later repaired. In figure 5.25 we see an indefinite pitch instrument, the taiko drum. Figure 5.26 shows one set of the vibrational patterns of a drum.

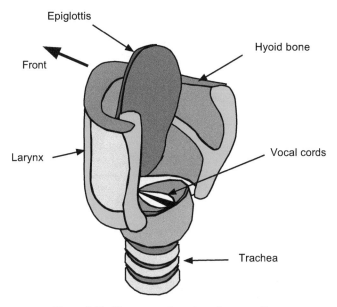

Figure 5.22. Human vocal cords and surroundings.

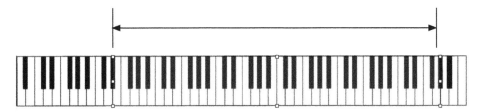

Figure 5.23. The vocal range of singer Mariah Carey of over five octaves is shown in blue and with brackets on the piano keyboard.

5.4 Resonance

Resonance is a phenomenon that is very important to many areas, including the production of musical sounds. Large scale oscillations can occur under appropriate conditions, as seen in figure 5.27: a large steel bridge was destroyed by resonant vibrations.

Resonance requires a mechanical system with a restoring force. When the system is disturbed or distorted, the restoring force brings the system back to its original condition. An example is Alice sitting on a swing, figure 5.28. If she is just sitting still, as in position A, there will be no movement, figure 5.28(a). When someone pulls the swing back to position B and releases it, the swing moves forward to C and then returns to B, figure 5.28(b). Here gravity plays the role of the restoring force. The swing takes a time T, called the period, for one complete cycle to occur: B to C and back to B. The frequency f of the system is the number of cycles in a unit of time—the inverse of the period. In the case of the swing, a type of pendulum, the frequency

Figure 5.24. A definite pitch percussion instrument: the Liberty Bell. National Park Service Photograph.

depends on the square root of the length, L. Note that this means that the natural frequency of a swing or pendulum does not depend on the mass of Alice, or on how far back the swing is pulled before it is released—only on its length.

A second requirement for resonance is an input of energy of a particular frequency; as a matter of fact, the necessary frequency is the natural frequency of the system. For the swing, Alice kicks back when she is at position B. This stores up increasing amounts of energy in the system and the swing goes higher and higher.

Another example of a resonant system is the trampoline. Here the elastic cords provide the restoring force and the trampoliner can reach dangerous heights by timing their jumps appropriately.

We mention resonance here because of its importance in coupling the vibrations of a sound to the surrounding air, in particular in the sound boxes of the stringed instruments. The air and sound box form a resonant cavity and the sound is amplified greatly. The oven part of a microwave oven, where the food is placed, is also a resonant cavity. Standing waves of the closed pipe type are formed. At the nodes, where the amplitude is at a minimum, cold spots occur, and if the food is not rotated during cooking, under-cooked spots result.

In conclusion, sound in general and music in particular play an important role in our everyday lives; now you may know a little more about some ways that we produce sounds.

Figure 5.25. An indefinite pitch percussion instrument: the taiko drum. Photo credit: http://en.wikipedia.org/wiki/File:Traditional-taikodrum-may2011.ogv. This image has been obtained by the author from the Wikimedia website where it was made available by Nesnad under a CC BY-SA 3.0 licence. It is included within this article on that basis. It is attributed to Nesnad.

5.5 The loudest sounds

Before we can discuss the loudest sounds around, we need to consider the origins of these sounds. First, there are loud sounds due to non-living sources, such as volcanoes and meteorites. Second, there are loud sounds of living things: animal sounds. Even though plants do produce sounds on their own, the sounds are very low in volume. Lastly, we recognize that some of the loudest sounds are generated by humans.

5.5.1 The loudest natural sounds due to non-living sources

The Earth, 4.5 million years ago, collided with a comet or meteor half its size. This event caused the biggest single atmospheric disturbance (and hence the loudest sound) ever. The result was our moon; the Earth barely escaped total destruction.

More recently, a meteor or comet landed about 66 million years ago near what is now the town of Chicxulub on the Yucatan peninsula of Mexico, figure 5.29. Its sound wave could have been heard thousands of km away, but not by humans, since they weren't

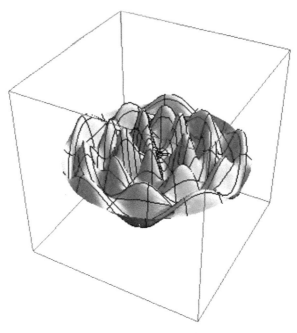

Figure 5.26. Vibrational patterns of a drum head. 'Normal Modes of a Circular Drum Head' by Adam Smith from the Wolfram Demonstrations project http://demonstrations.wolfram.com/NormalModesOfACircularDrumHead.

Figure 5.27. The Tacoma Narrows Bridge was opened for traffic on July 1, 1940, and collapsed on November 7, 1940. Photo credit: Library of Congress (LC-USZ62-46682).

(a)   (b)

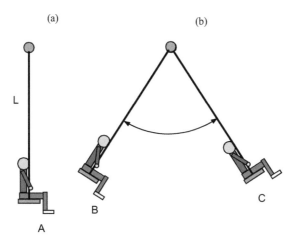

Figure 5.28. The swing (a) at rest and (b) swinging. The swing is a resonant system.

Figure 5.29. The location of the Chicxulub crater in Mexico dating back 66 million years. The impact and its aftereffects are thought to have resulted in the demise of the dinosaurs.

around yet. The impacting object was about 50 km (30 mi) in diameter and had a mass of around 4×10^{17} kg or 4×10^{14} tons. One result was the extinction of the dinosaurs.

Other extraterrestrial objects have impacted the Earth more recently. One such occurrence was the Tunguska Event of 1908, another is the meteor that exploded over Chelyabinsk, Russia, on February 15, 2013.

Both of these were very loud, but nothing like the volcanic Krakatoa eruption of 1883, figure 5.30. The resulting sound of about 310 dB is generally accepted as the loudest natural sound ever produced in modern times. Occurring in the Sunda Strait of Indonesia, 68% of the Island of Krakatoa simply disappeared; 36 000 deaths were recorded and the blast was felt as far away as New York City, about 16 000 km or 10 000 mi away.

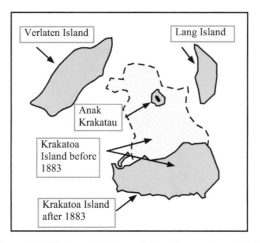

Figure 5.30. Krakatoa before and after the eruption of 1883.

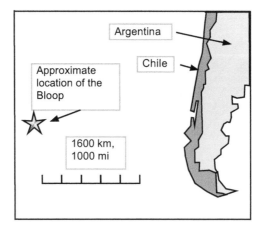

Figure 5.31. 'The Bloop's' location off the coast of Chile in the Pacific Ocean.

Underwater, where sound travels over 4 times faster than in air, '*The Bloop*' is the record holder for the loudest natural sound. In 1997, scientists from the United States National Oceanic and Atmospheric Administration (NOAA) recorded a very high intensity (172 dB) sound of very low frequency; it lasted several dozen seconds. It was subsequently named 'The Bloop,' and its source was pinpointed off the coast of Chile. The Bloop was detected by several separate underwater hydrophones 5000 km (3000 mi) apart. Never again recorded, the origin of 'The Bloop' remained a deep mystery until 2008, when NOAA decided that it was a case of an icequake or iceberg calving (figure 5.31).

5.5.2 The loudest animals

5.5.2.1 Land animals

The loudest land animals that we can hear today are howler monkeys and elephants. The forests and jungles of South and Central America provide the habitat of the howler monkey. A rather large animal (up to 1.2 m (4 ft) in length and 10 kg (22 lb), the howler can produce sounds with a volume of up to 140 dB. Figure 5.32 shows a howler monkey in action. They can be heard up to 4.8 km (3 mi) away.

Elephants also can be very loud. They are capable of producing sounds of over 110 dB and they can detect their own low frequency sounds at up 4 km (2.5 mi). Figure 5.33 shows a diagram of an elephant's head and the main parts of its sound generating

Figure 5.32. Howler monkey howling. This image has been obtained by the author from the Wikimedia website where it was made available by Steve under a CC BY-SA 2.0 licence. It is included within this article on that basis. It is attributed to Steve.

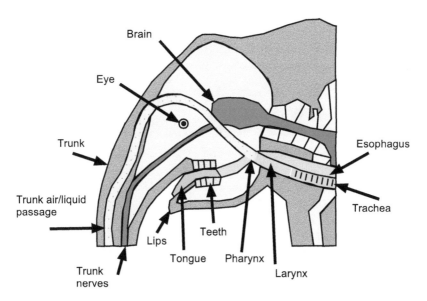

Figure 5.33. Cross-section diagram of an elephant head.

Figure 5.34. Trumpeting elephant. Credit: Claudia de Mayo

system, including its larynx, or voice box. Figure 5.34 is a photo of a bull elephant trumpeting. If you see this live and up close, beware: the elephant is about to charge!

5.5.2.2 Aquatic animals

There are two candidates for the loudest aquatic animals: the *mantis shrimp* and the *sperm whale*, both are very interesting creatures—one small and the other quite large. We have already seen the mantis shrimp's amazing eyes (page 3-20).

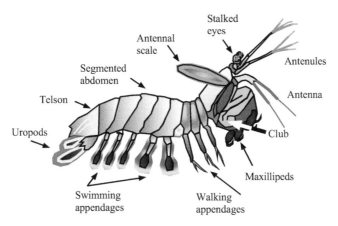

Figure 5.35. Parts of the peacock mantis shrimp, *Odontodactylus scyllarus*.

Mantis shrimps are crustaceans, but not actually members of the shrimp family. Mantis shrimps come in two types: the smashers and the spearers. The smashers have club-like appendages capable of breaking thick glass and of producing some of the loudest sounds of the ocean, figure 5.35. The smasher's mechanism and parts are given in figure 5.36(a) and (b). The spearer version has spears with sharp spikes and barbs, figure 5.36(c). Normally folded, the spears are as long as the mantis shrimp's body when extended. The similarity to the arms of the praying mantis insect give the mantis shrimp its name.

The mantis shrimp has powerful muscles in its clubs. These muscles usually keep the clubs folded up under tremendous tension. Other muscles keep the clubs from extending. Similar to this is the cocked finger when it is about to flick, figure 5.36(a) and (b). When the cocked mantis muscles are released, the clubs move forward with tremendous speed, as does the flicking finger. The club acts with accelerations of over $100\,000$ m s^{-2} ($10\,000$ times the acceleration due to gravity), and with speeds of up to 80 km s^{-1} (50 mph). The resulting impacts can cause cavitation bubbles in the surrounding water. When the bubbles collapse, they release enough energy to kill or stun nearby animals, making an easy meal for the mantis. The huge energy release also can result in the emission of visible light—sonoluminescence. The sound level generated is up to 210 dB.

Another loud water animal is the Sperm Whale, *Physeter macrocephalus*. The slightly larger Blue Whales and Fin Whales are both baleen whales; they use filters to help them eat small crustaceans called krill. The sperm whale, however, is a toothed whale and feeds mainly on giant or colossal squid, the largest invertebrate on Earth at 10 m (33 ft) long. To assist in finding its prey, the sperm whale has a powerful and sophisticated echo location (sonar) sound system. The whale communicates with its companions with calls of up to 188 dB and its echo location clicks are an amazing 230 dB.

The sperm whale is an animal of many superlatives; of a length of over 20 m (66 ft) and a mass of 80 000 kg (176 000 lb, 88 US tons), figure 5.37. Its brain at 8 kg (17 lb) is the largest known of any animal, living or extinct, and it is 5 times heavier than the human brain. The sperm whale also has the longest intestinal system at 300 m (328 yards).

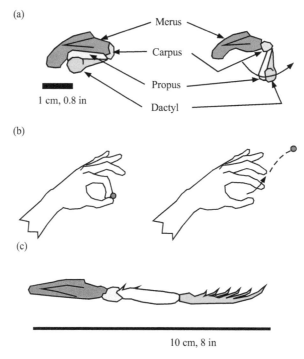

Figure 5.36. (a) The club mechanism of the mantis shrimp. (b) Finger flicking a pea, in a manner similar to the mantis shrimp striking an object. (c) The spear of the spear version of the mantis shrimp.

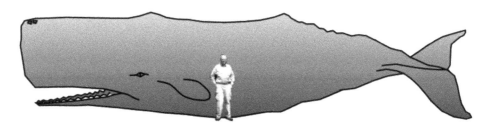

Figure 5.37. A photo of the author next to a drawing of a sperm whale.

The sperm whale gets its name from the oil-like spermaceti in the upper part of its head, figure 5.38. In olden days, whalers like the fictional Captain Ahab sought out and killed whales such as Moby Dick.

The whalers were after the whales' spermaceti, rendering it into valuable sperm oil for lamps and other uses such as lubricants. The whales' blubber was also made into whale oil, which was less valuable than sperm oil.

The sound emitting/receiving process is straightforward and highly effective. Starting on the left of the diagram, the blowhole is located off-center to the left of the head's center line. There are two separate nostrils. The left nostril is connected directly to the blowhole. The right nostril goes through the 'monkey lips' which function as vocal cords. The air trapped in the lungs is released through the right

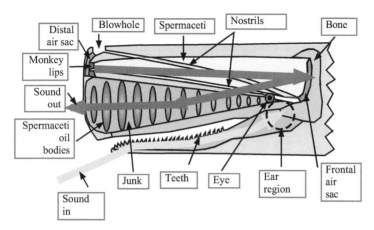

Figure 5.38. Diagram of the parts of the sperm whale sound system.

Figure 5.39. Drawing of the 'Tsar Bomba,' along with a technician in a hazmat suit, approximately to scale. The bomb was about 8 m (26 ft) in length and 2.1 m (7 ft) in diameter. Its mass was around 12 000 kg (26 000 lb). The USSR detonated the Tsar Bomba over the Barents Sea on October 30, 1961. It was found to be too devastating to use, even in warfare.

nostril into the monkey lips where sound is produced. The generated sounds reflect off the front (distal) air sac and travel backwards through the spermaceti. Reflecting off the frontal air sac, the sounds travel forward again through the junk and its disk-shaped spermaceti oil bodies. The sounds are shaped and directed according to their uses: communications with other whales.

5.5.3 The loudest sounds produced by humans

In the category of the loudest sounds produced by humans, the champion is the Tsar Bomba, figure 5.39, a hydrogen bomb. On October 30, 1961, the USSR detonated it in the air above the Barents Sea; it was the most powerful device ever detonated by humans, before or since. With an energy of about 200 000 TJ, it was over 3000 times the energy of the A-bomb that destroyed Hiroshima, Japan, on August 6, 1945. The Tsar

Figure 5.40. A Space Shuttle launch. Note the steam and the elevated water tank for holding the water. Credit: NASA/JSC.

Bomba's shock wave blast broke windows over 800 km (500 mi) away. Its loudness is estimated to have been 224 dB, loud enough to kill anything too close. After an analysis of its effects, the USSR decided that the Tsar Bomba was too big and destructive to use, even in war. Andrei Dmitrievich Sakharov (1921–1989) was one of Tsar Bomba's key developers. Upon considering its power, he became one of the main supporters of the Limited Nuclear Test Ban Treaty of 1963. He was awarded the 1975 Nobel Peace Prize.

Another loud sound occurred when a large payload, such as the Space Shuttle, launched from the Kennedy Space Center. The rocket's engines had a sound output of about 202 dB, enough to injure the crew and damage the rocket's hardware if unabated. The scientists and engineers developed a sound suppression system. It flooded the launch pad with 300 000 gallons (1.1 M l) of water to a depth of 15 cm (6 in.) of water during launch. An elevated water tank supplied the fresh water used. The 'smoke' seen during a launch is mostly water-steam formed when the heat of the rocket engines hits the suppression water, figure 5.40.

This brings us to the end of chapter 5; the next chapter will explore new aspects of vision such as augmented and synthetic vision systems.

IOP Publishing

The Everyday Physics of Hearing and Vision (Second Edition)

Benjamin de Mayo

Chapter 6

Vision and the perception of light and color

In this chapter, we will investigate some of the aspects of our perception of light: light intensity levels, blindness prevention, cataracts, color, paint and color blindness.

6.1 Light intensity

Our eyes can detect a tremendous range of light intensities. Photometers are used to measure light levels; almost all are based on the photocell, a transducer that changes light energy into electric energy in the form of a voltage, figure 6.1. Light levels are important to photography and to workspace efficiency and safety. Cameras and smart phones today have built-in photometers for measuring light levels. Light bulbs are rated according to how much light they produce.

The measurement units of light intensity are complicated. We measure luminous intensity in lumens. One foot candle, figure 6.2, is equal to one lumen per square foot, which makes it a measure of luminance, the luminous intensity per area.

Light bulbs are rated according to their light production and their energy consumption. Figure 6.3 compares incandescent light bulbs and light emitting diodes (LEDs). Figure 6.4 shows the recommended light levels for various activities.

The response of our eyes to light levels is strikingly similar to the response of our ears to sound levels. *Fechner's law* says that the subjective sensation of brightness is proportional to the logarithm of the intensity measured in decilams:

$$B = 10 \times \log(I_1/I_2),$$

where B is the brightness in decimals and I_1 and I_2 are the two intensities being compared. Figure 6.5 compares different light levels and different sound levels. Both responses are logarithmic.

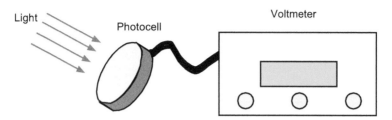

Figure 6.1. The photometer measures light intensities. Now available as a free smart phone app.

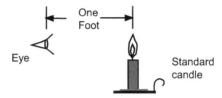

Figure 6.2. The foot candle, a measure of illuminance. A completely reflecting surface illuminated by one foot candle has a luminance of one foot lambert.

a) LED bulbs

Input Power, Watts	Lumens Produced	Lumens/ Watt
6	550	91.7
12	1000	83.3
150	15,600	104.0

b) Incandescent bulbs

Input Power, Watts	Lumens Produced	Lumens/ Watt
25	260	10.4
100	1,580	15.8
1000	20,700	20.7

Figure 6.3. LED bulbs (a) compared to incandescent bulbs (b).

Area	Nominal Illumination Level in Lumens/Square Meter (
Normal work station space, open or closed offices	500
Conference Rooms	300
Internal Corridors	200
Auditoria	150-200
Public Areas: stairwells, lobbies, corridors	200
Toilets, locker rooms, janitors' closets, shops, docks	200
Dining areas	150-200
Kitchens, gymnasia, child care centers	500
Parking garage	50

* 1 lux = 1 lumen per square meter = 11 footcandles

Figure 6.4. Recommended lighting levels for various areas.

6.2 Blindness prevention

The leading cause of blindness in people younger than 21 years of age is injury to the eye. This means that we should always be aware of the potential for accidents and wherever possible use safety glasses. A good substitute for safety glasses is regular glasses whose lenses have been heat treated to improve their shatter performance.

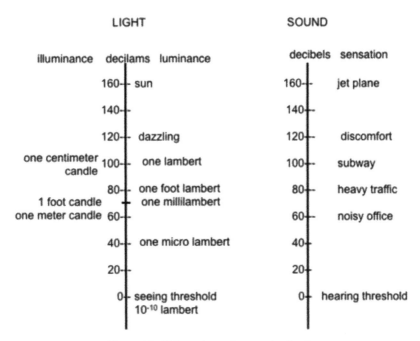

Figure 6.5. Light and sound perception levels.

It costs only a small amount more at the time of purchase to obtain this extra feature for your glasses. The fact that the improved lenses may save your eyes in case of an accident makes this extra feature well worth the money. Plastic lenses do not need this extra treatment; they are already shatter resistant. You should also be aware of the fact that contact lenses offer little or no protection from injury.

After the age of 21, the leading causes of blindness are diabetes, macular disease and glaucoma. Glaucoma is a pressure buildup inside the eyeball that is sometimes brought on by diabetes. A cataract is a clouding of the lens. Cataracts increase in frequency with advancing age and when severe are treated by replacing the lens. The reason for the development of cataracts is shown in figures 6.6 and 6.7; figure 6.8 presents a diagram of a replacement lens. In the embryonic development of the eye, a stalk (pink) emerges from the proto-brain, epithelial skin cells (green) form a cup-like structure; the surface cells now face inside the cavity. The brain cell stalk also forms a cup-like structure, which will become the eyeball. This is the origin of the inside-out nature of the retina, in which light has to go through the ganglion layer before it reaches the retinal cones and rods. There is a direct connection between the eyeball and brain. As the skin-cup closes, the outer layer becomes the cornea. The horizontal epithelial skin cells (blue) form inside the proto-lens, figure 6.6(d)–(f). These interior lens cells are arranged parallel to the direction of the incoming light, thus enhancing the transparency of the lens. Figure 6.6(f) shows the appearance of the ciliary fibers (orange).

The skin cells of the lens grow toward the center of the lens. Like normal skin cells, they die. But instead of being sloughed off, as in the case of the dead skin cells

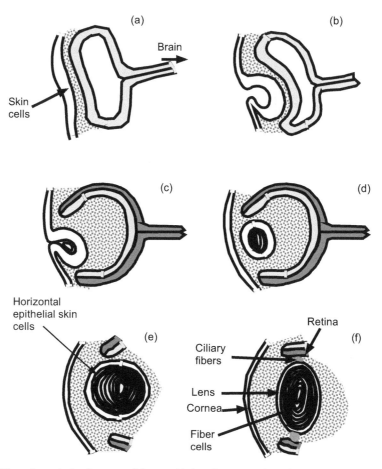

Figure 6.6. The embryonic development of the eye. (a) A stalk emerges from the proto-brain and forms a cup. (b) The skin cells (green) form a pouch and the brain cells (red) form a cavity as well. (c)–(e) The skin cell pouch closes and horizontal epithelial cells (blue) develop. (f) The ciliary fibers (orange) form. Light is entering from the left.

on your arm, for example, the lens' dead skin cells accumulate in the middle, eventually becoming cataracts. Figure 6.7 shows a sectioned lens.

To address the cataract problem, the defective lens is dissolved with ultrasonic waves, sucked out of the eyeball and then replaced by an intraocular plastic lens. Its 'wings' fix it in the eyeball, figure 6.8. A special design provides both near and distant correction for the eye.

Great progress has been made in the development of artificial cochleas for the hearing impaired. Although there have been significant recent advances, progress is more limited for retinal implants to assist in vision; see section 6.5 later in this chapter. The vision health professionals include the ophthalmologist: a medical doctor who has completed four years of medical school plus advanced training in vision and who can prescribe drugs and operate on eyeballs. The optometrist, titled OD, has completed four years of optometry school and can prescribe certain drugs.

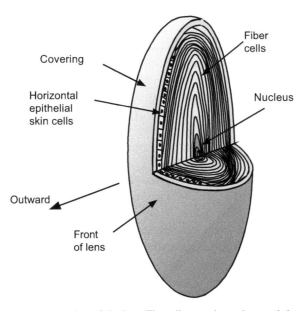

Figure 6.7. A cross-section of the lens. The cells grow inward toward the nucleus.

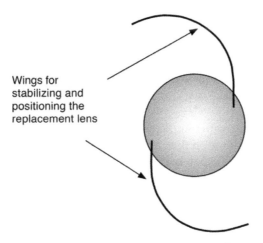

Figure 6.8. An intraocular replacement lens to correct for cataracts.

The optician makes the glasses that the ophthalmologist and optometrist order for you.

Unfortunately for us, our bodies were not designed to last much past 40 years. Once you hit this age, various things start to go; if you are lucky, your systems will taper off slowly. One of the first symptoms of aging is that your eye lens will become less flexible. You will become farsighted and need reading glasses or bifocals if you want to read the fine print. In spite of this, with a little care and a little luck, your eyes should serve you well for a (hopefully) long life.

6.3 Color

The existence and nature of color has always fascinated mankind. Even in the most primitive of societies, color plays a large role in such areas as religion, art and trading. Color is also vitally important in the lives of many types of animals and plants. Figure 6.9 is a table of the wavelengths for the seven basic color designations.

By the 1600s, the scientific study of color had resulted in the consensus belief that light itself has no color and that the property of color belongs to the objects being illuminated. In other words, light merely makes the objects visible. Descartes, the French philosopher, correctly proposed in 1630 that the color of an object is attributed to a change in the light reflected from the object.

The perception of color is a psychological phenomenon; it depends on the energy of the physical stimulus and the response to the stimulus. The color of an object depends on the wavelength of the light striking it, the light's intensity, the color of any surrounding objects, the length of time that the object is observed, substances in the path of the light, the psychological state of the observer and possibly other factors. The three psychological aspects of color are *hue*, *saturation* and *brightness*, none of which are directly measurable. The associated psychophysical variables (which can be measured) are *dominant wavelength*, *purity* and *luminance*, respectively.

Hue is the reaction of color vision to the different parts of the spectrum: red, green, blue, yellow, etc. *Spectral hues* can be matched with light of one wavelength; about 200 of these are distinguishable, one for every 2 nm in the visible spectrum. The *non-spectral hues* must have at least two wavelengths for a match; purple is an example. Saturation refers to the degree of hue in a color: pale or rich, weak or strong. Pink is low-saturation red; scarlet is high-saturation red (figure 6.10).

Purity is a psychophysical quality of color. For example, scarlet is 100% pure; pink has a low purity, since it is white mixed with red, figure 6.11. Because of the eye's differing sensitivity to the different areas of the spectrum, colors of the same purity do not all have the same saturation. For example, pure yellow is much less saturated than pure violet. This means that it is much easier to produce high-saturation (bright) yellow than high-saturation violet. This has important implications for artists when

Color	Wavelength, nanometers	Range, nanometers
Violet	350-400	50
Indigo	400-450	50
Blue	450-500	50
Green	500-550	50
Yellow	550-620	70
Orange	620-670	50
Red	670-750	50

Figure 6.9. The wavelengths and ranges of the basic colors.

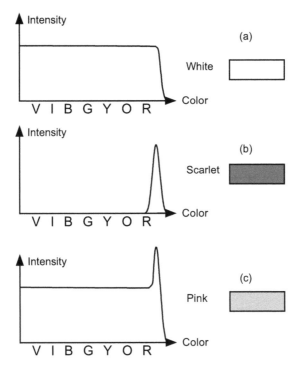

Figure 6.10. (a) White is an even mixture of all wavelengths. (b) Scarlet is 100% red (high purity). (c) Pink is red with other wavelengths (low purity).

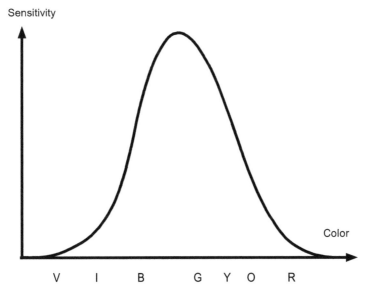

Figure 6.11. The sensitivity of the human eye. The eye is more sensitive to the middle of the spectrum than the edges. This affects color perception.

they mix colors. Luminance also affects saturation. Blue and red appear more saturated at low luminance than yellow, which needs higher luminance to achieve the same saturation.

The 10 million retinal cones in each of our eyes are our color receptors. There are three different types of cones, 'red', 'blue' and 'green', figure 6.12. Each type of cone is sensitive to a range of wavelengths, but its sensitivity is centered on the color of its label. For example, the 'red' receptors can detect wavelengths from violet to red, but

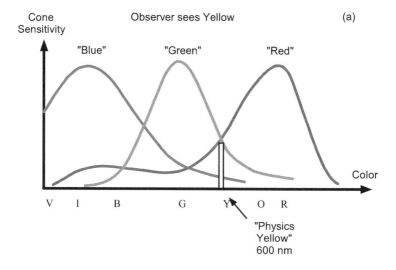

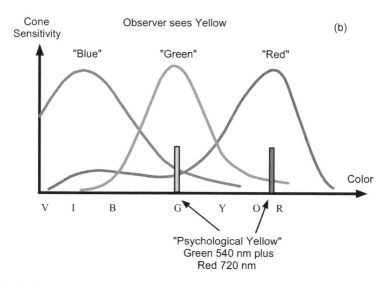

Figure 6.12. The three receptors 'blue', 'green' and 'red', enable us to see all colors. Compare (a) 'physics yellow' (yellow light of 600 nm wavelength) and (b) 'psychological yellow' (an equal combination of green (540 nm) and red (720 nm) light).

are most sensitive to red light. All the colors of the rainbow can be perceived with just these three types of cones.

The human eye (unlike the ear) cannot break a wave down into its component wavelengths. For instance, there are two colors 'yellow'. Figure 6.12(a) shows how exposing the eye to yellow light of 600 nm (we call this *'physical yellow'*) will produce the sensation of yellow by equally stimulating the 'red' and 'green' cones. We can also produce the sensation of yellow by sending equal amounts of red and green light into the eye, figure 6.12(b); we call this *'psychological yellow'*. The eye cannot distinguish the two yellows; both look identical.

'White' can result from an even mixture of all wavelengths or from an equal mixture of red, green and blue. Children know that the primary colors in a crayon set are blue, red and yellow. Green is made by coloring blue and then yellow. As we will see, this nomenclature is wrong. Figure 6.13(a) shows physics white: an even

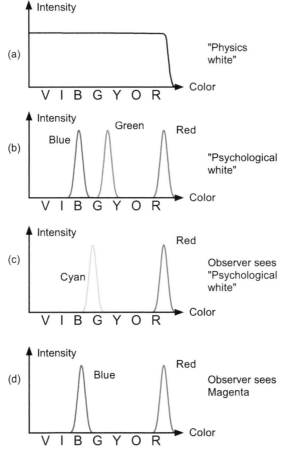

Figure 6.13. Some colors. (a) Physics white: all wavelengths. (b) Psychological white: only red, blue and green. (c) Cyan: blue plus green; added to red cyan yields psychological white. (d) Red plus blue yields magenta, a non-spectral color.

mixture of all of the visible light wavelengths. Psychological white results when only red, green and blue, the additive primary colors, are present, figure 6.13(b). When red is added to cyan (blue plus green) we also obtain the sensation of white, figure 6.13(c). The color magenta results from the addition of red plus blue, figure 6.13(d). Magenta is called a *non-spectral color*, it is not in the ROY G BIV spectrum of white light. Other non-spectral colors are cyan and purple. These are not found in the color spectrum as a single wavelength.

Any two colors that can be added together to obtain white are called complementary colors. A helpful visual aid is the color triangle, figure 6.14. The complementary colors are arranged opposite each other on the color triangle.

For example, blue and yellow are complementary colors; when added, they yield white. Figure 6.15 also shows this with the color wheels. Adding two of the additive primaries yields one of the subtractive primaries; adding all three yields white: the center of the circles. On the other hand, the subtractive color wheels show that

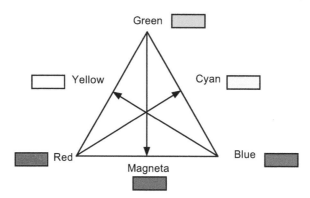

Figure 6.14. The color triangle. At the points are the additive color primaries; the subtractive primaries are on the sides. Any color can be made by adding the additive primaries and conversely any color can be made by subtracting a combination of the subtractive primaries from white light.

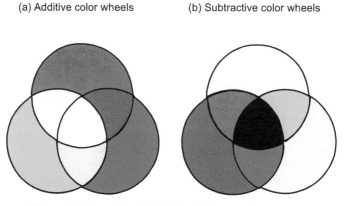

Figure 6.15. The color wheels: (a) additive and (b) subtractive.

subtracting any two of the subtractive primaries from white light will result in one of the additive primaries; subtracting all three yields black, the center of the wheels. Figure 6.16 shows how red, green and blue spotlights can overlap to show the complementary colors. For example, where the red and green spotlights overlap, one sees yellow, even though clearly there is no yellow light present.

A phenomenon called *afterimage* can be demonstrated. Stare at the white dot of figure 6.17 for 60 s and then look slightly away; a red, white and blue flag will appear. The colors red, white and blue are the complementary colors of cyan, black and yellow.

Examples of additive color mixing to make orange and pink are shown in figure 6.18. In the first example, green plus red yields yellow; however, if we add a little extra red, we obtain yellow plus red, that is, orange. In the case of pink, the blue, green and red yield white; but, we've added a little extra red again, making our white a little redder, namely pink.

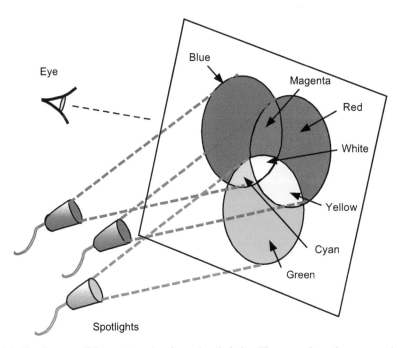

Figure 6.16. The three spotlights cast overlapping colored circles. The area where the green and red lights overlap appears yellow, but no yellow light is present.

Figure 6.17. Use this to observe a red, white and blue afterimage. Stare at the white dot for 60 s and look to the side.

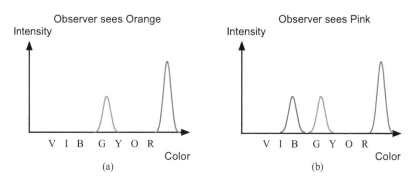

Figure 6.18. How to make (a) orange and (b) pink using the additive primaries.

Television screens and the Sunday comic pages in newspapers use red, green and blue dots to achieve additive color mixing. Previously, a close-up view of a television screen would have revealed that there are only red, blue and green dots. The dots were more visible if you flicked some small water droplets onto the screen. Because of surface tension, the water beaded up into tiny magnifying glasses. The dots are more difficult to see now because of high resolution displays.

Subtractive color mixing uses the subtractive primary colors, yellow, magenta and cyan. Any color can be made by subtracting combinations of the three subtractive primaries out of white light. Figure 6.19 shows how yellow light can be can be made subtractively by using a filter that removes all of the wavelengths except yellow. '*Physics yellow*' results. Using this method, we would need an appropriate filter for every color that we intend to produce. Or, we could use combinations of only the three subtractive primary filters to make any color. Figure 6.19 also shows how a subtractive primary filter that removes only blue from red–green–blue white light will let red and green pass through, giving us 'psychological yellow'.

Figure 6.20 shows how to make orange and red subtractively. Note that the weak magenta filter lets a little green through, thus producing orange.

In figure 6.21, we see how to make pink with the subtractive primaries.

Most of the colors that we encounter in day-to-day life are produced with dyes. For a red shirt, when white light falls on the shirt the dye in the cloth absorbs all the colors except red, which is reflected.

Paint is made by combining a pigment in a translucent binder and applying it to a base—the object to be painted, figure 6.22.

You could attempt to find a suitable pigment for every imaginable color or you could use the three subtractive primaries to produce any color. Now we can solve the crayon problem. To obtain green by subtractive color mixing, we combine yellow and cyan; thus the 'blue' in the crayon set is really 'cyan'. This process must take into consideration the relation between hue and saturation. At least one manufacturer of artists' oil and latex paints has tried to simplify paint mixing by assigning numbers for hue and saturation to the different paints.

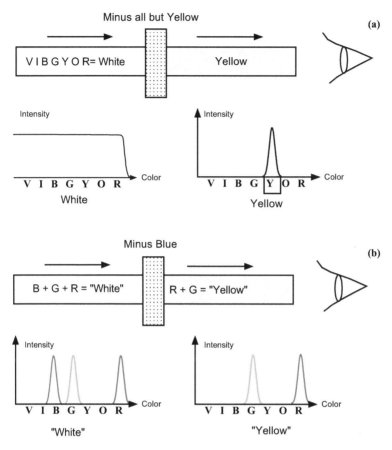

Figure 6.19. How to make yellow light subtractively in two different ways. (a) Use a filter that removes everything but yellow or (b) use a subtractive primary filter that passes only red and green.

6.4 Other aspects of vision

6.4.1 Color blindness

About 8% of human males have difficulty discerning colors, whereas only 1% of females have a color deficiency. In most cases, these deficiencies are related to one of the three types of cones. For example, red–green color blindness is due to a problem with the 'blue' receptors. Figure 6.23 shows a typical color blindness test figure. I hope that you can find the '74' in it. Entire color blindness evaluations can be found on-line. A person with normal color vision is called a *trichromat*: she can match any color with a combination of three colored lights. There are three types of color vision deficiency. Anomalous trichromats are poor at mixing colors, especially green. *Dichromats* can match any color to their own satisfaction with only two colors instead of three. Whereas the results will look okay to the dichromat, they will look discordant to a person with normal color vision. The *monochromat* sees no color; everything looks like a black and white photograph.

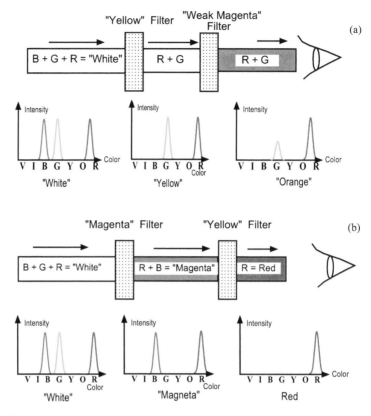

Figure 6.20. How to use the subtractive primaries to make (a) orange and (b) red.

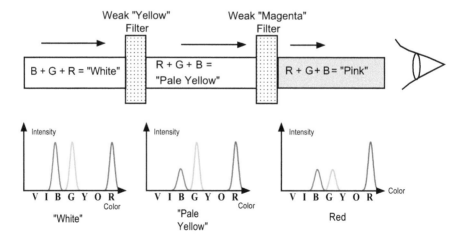

Figure 6.21. How to use the subtractive primaries to make pink.

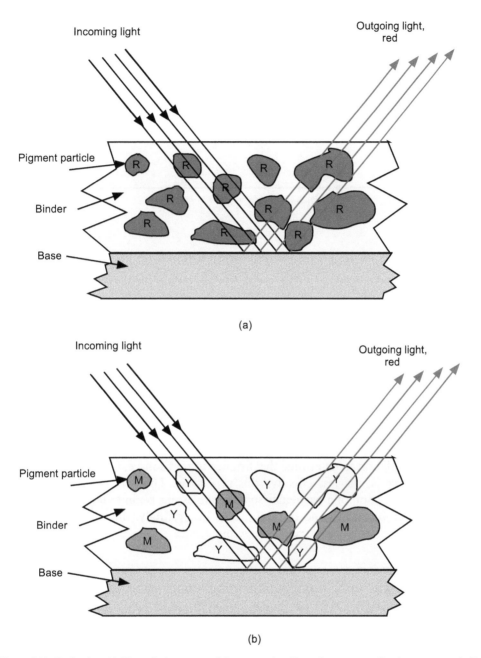

Figure 6.22. Red paint. (a) The red pigment particles act as tiny filters that remove all colors except red. (b) Using the subtractive primaries, the magenta pigment particles pass the red and blue, stopping the green light. The yellow ones pass the red and green, stopping the blue; red light remains.

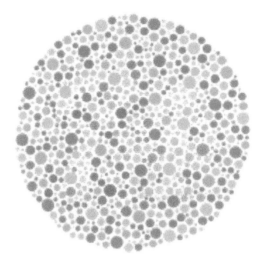

Figure 6.23. Color blindness test figure. Do you see a '74'? This image of Ishihara Plate 9 has been obtained by the author from the Wikimedia website https://commons.wikimedia.org/wiki/File:Ishihara_9.png, where it is stated to have been released into the public domain. It is included within this article on that basis.

6.4.2 Stereo

Having two ears and two eyes provides a form of insurance in case of the loss of one. Dual ears and eyes also give us the ability to determine the distance or direction of a sound source or visual object. These abilities are of great utility and have helped us to survive as a species. Today stereo audio and 3D movies add greatly to our enjoyment of entertainment. Due to the distance between our eyes, each focuses in a slightly different direction toward the source of interest. Our powerful brains do an automatic calculation to determine the distance to that source. Of course, the further away the source is, the less effective this process is. The very slight difference in timing of the detection of a particular sound enables our brain to tell the direction of the sound, even behind us and through walls. Interestingly, this does not work too well for sounds originating directly above our heads.

6.4.3 Optical illusions

The eye sends an intelligent message, not a photograph-like image, to the brain. An easy way to observe this is by looking at optical illusions, figure 6.24. You might say that the eye is tricked into seeing something that is clearly not right. But this is the price we pay; otherwise the brain would be completely overwhelmed by the vast amount of information gathered every second by each eye.

6.5 New developments

6.5.1 Retinal implants

The same technology that brought you the smart phone and the in-ear wireless hearing aid (page 2-16) is at work in your eye. For example, an Israeli company, Nano Retina (www.nano-retina.com) offers a wireless retinal implant. It uses micro-miniaturization,

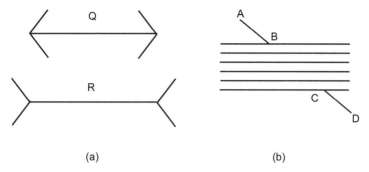

(a)   (b)

Figure 6.24. Two simple optical illusions. (a) Lines Q and R are of equal length. (b) Lines AB and CD go straight through the parallel lines.

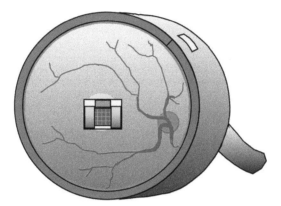

Figure 6.25. Implant mounted directly on the retina.

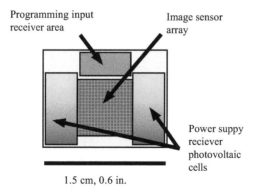

Figure 6.26. Diagram showing the parts of the implant.

advanced programming, nano-technology and modern battery advances. The device is pressed into the retina in the back of the eye, figure 6.26. (Figures 6.25–6.30 are drawings adapted from Nano Retina's web site; the dimensions are approximate, and the enhanced colors are pedagogical.)

The implant is about the size of a thick postage stamp and covers much of the fovea. Its several parts are shown in figure 6.26. The *image sensor array* contains hundreds of individual units similar to those in smart phone cameras. The *programming input receiver* allows the accompanying eyeglasses, figure 6.27, to communicate wirelessly with the implant and to adjust the electronics. The glasses also send power wirelessly to the *power supply receiver* area of the implant. Photovoltaic cells similar to solar cells pick up an infrared beam from the glasses and recharge the implant's battery. Figure 6.28 is a close-up diagram of the nano-spikes, which penetrate the retina all the way down to the sensory cells.

Light enters the eye, passes through the lens and iris and is focused onto the image array. The individual sensors of the array send signals to the micro-electronics layer. In this layer, the signals are analyzed and converted to voltages appropriate for the actual retinal cells. The signals are then passed into the retina via the nano-spikes and sent eventually to the brain.

Figure 6.29 shows in diagrammatic detail a cross-section of the implant attached to the retina.

This device is quite promising. Instead of ten or twenty artificial sensors for the retina, as current devices provide, thousands will be in use for each retina.

6.5.2 Gene therapy for blindness

Recently, gene modification therapy has been applied to an inherited form of blindness. The technique uses CRISPR, an acronym for Clustered Regularly

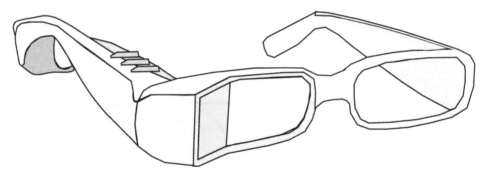

Figure 6.27. Diagram of the glasses for the retinal implant system.

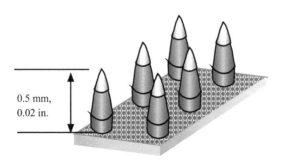

0.5 mm, 0.02 in.

Figure 6.28. Nano-sized spikes for transmitting signals into the retina.

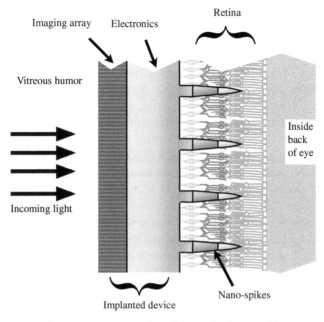

Figure 6.29. Diagrammatic cross-section of the device implanted into the retina.

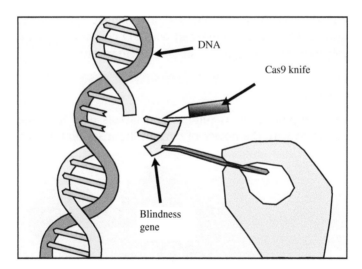

Figure 6.30. In the CRISPR technique, a defective gene is removed, repaired and reinserted into the DNA strand.

Interspaced Short Palindromic Repeats and the enzyme Cas9 (CRISPR **as**sociated protein **9**). The enzyme acts as a molecular knife and cuts the defective gene from the appropriate DNA strand. The gene is repaired and reinserted into the DNA strand. This powerful technique lends optimism to the search for the treatment of many human ills. Figure 6.30 shows how CRISPR works.

6.5.3 Augmented vision systems

Just as with hearing aids, humans have devised external ways to augment our vision. Eyeglasses, contact lenses, telescopes, binoculars and microscopes are examples of non-electronic instruments that improve our visual powers. Micro-electronic improvements have made possible an ever-increasing array of visual instruments. The gyro-stabilized binocular is a good example of this trend, see figure 6.31.

Improvements in our ability to see at night date back to regular binoculars. Besides magnifying an object, binoculars provide much brighter images. Since their apertures are much larger than our own pupils, the binoculars gather much more light. The large increase in the amount of light gathered results in a much brighter image for us.

An interesting example of available night vision devices is given in figure 6.32. For a modest price, you get not only photo and video modes but also an LCD (liquid crystal display). An integral infrared source supplies seven levels of radiation as needed.

LiDAR stands for Light Distance and Ranging. Previously a single LiDAR unit would cost at least several tens of thousands of euros and was quite large in size, often requiring a separate cooling unit. Airplane borne units have revealed such items as temples in hidden in jungle foliage, figure 6.33.

Now you can get your own inexpenive add-on LiDAR kit for auto. LiDAR is now standard on some car models; its display shows up on the auto's dashboard screen. LiDAR is a viable candidate for driverless vehicle systems. Or you can buy an iMac Pro® with a LiDAR scanner included, figure 6.34. One of the potential uses of the iMac/LiDAR systems is for AR/VR (augmented reality/virtual reality). In AR, a real-world environment is enhanced with 3D computer-generated objects or perceptions. VR presents a completely simulated environment.

6.5.4 Synthetic vision

Today's desktop computers, smart phones and tablets now have incredible capabilities. When coupled with the power of current flight simulator software, you can

Figure 6.31. Gyro-stabilized binoculars. Credit: Canon.

Figure 6.32. (a) An example of inexpensive night vision binoculars with an LCD screen and an infrared light. Rexing Model B1, www.rexingusa.com. (a) Rear view, (b) front view.

Figure 6.33. LiDAR revealed the Grand Plaza of the Mayan city of Tikal in Guatemala even though it was heavily covered with jungle vegetation. Credit: Image created by Juan C Fernandez-Diaz from airborne lidar data collected by NCALM for the Pacunam Lidar Initiative (PLI).

have a startlingly realistic virtual flying experience. For example, you can 'fly' down the Grand Canyon and observe individual rocks and trees. If you select 'winter' as the time of your flight, each individual rock and tree has a layer of snow. For an autumn flight, the deciduous trees have fall foliage while the conifers are green. The

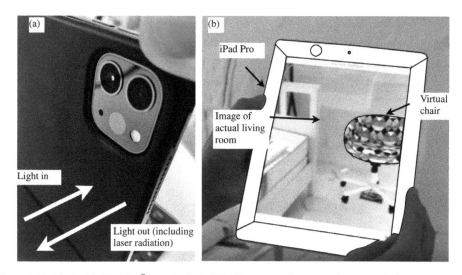

Figure 6.34. (a) Apple's iPad Pro® has a built-in LiDAR scanner. This image has been obtained by the author from the Wikimedia website where it was made available under a CC BY-SA 4.0 licence. It is attributed to KKPCW. (b) Example of a furniture possibility for a living room.

Figure 6.35. Flight simulator view from the cockpit of a plane landing, X-Plane® flight simulator.

computer calculates and shows on your screen in incredible detail what you would see if you were actually flying and looked out the window of your virtual airplane. When you glance down, the plane's instrument panel is also shown; the gauges' needles and displays all 'work' in real time.

Figure 6.35 shows the virtual view from the cockpit of an airplane coming in for a landing using the flight simulator, X-Plane®. The image seamlessly updates as the plane lands on the runway, under control of the virtual pilot. The user can not only select from dozens of planes of all types, but also the plane's location, and the

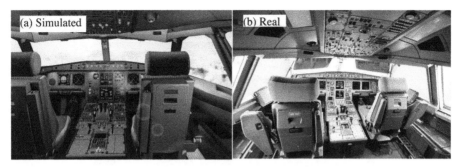

Figure 6.36. The cockpit of an airliner. (a) X Plane® simulated cockpit of an airliner courtesy of JARDesign. (b) Photo of a real A330 cockpit. This Airbus A330-200 D-ALPA image has been obtained by the author from the Wikimedia website where it was made available under a CC BY 3.0 licence. It is attributed to Ralf Roletschek.

Figure 6.37. The view of an actual airplane's instrument panel with synthetic vision installed. Note the clouds visible through the windshield and the synthetic vision displays. Copyright 2020 Garmin Ltd or its Subsidiaries. All Rights Reserved.

weather (selectable or real time), season of the year and the time of day. Figure 6.36 compares a simulated airliner cockpit and a photo of the real thing.

Synthetic vision is a recent development which incorporates these capabilities into an actual airplane, for a cost of course. Figure 6.37 shows the instrument panel of such a real airplane. The computer-generated view on the left is what would be visible out of the windshield, even though the real plane is flying in totally cloudy or dark weather. Also shown on the right display is a GPS map of the plane's actual location, giving such information as altitude, speed and the distance and direction to local airports. This built-in capability costs thousands of Euros, even on small private planes.

This capability is a feature on current flight simulators; figure 6.38 shows the two displays on a simulated cockpit.

There are now apps available for tablets and other similar devices for only dozens, not thousands, of Euros, figure 6.39.

So here's the bottom line: you can get the same capabilities in your €200 million ($230 million) Airbus 330 with synthetic vision as you've already had on your X-Plane® flight simulator. And for a small yearly fee, you can get a synthetic vision app for your laptop or tablet that you can use in your own airplane and that updates automatically.

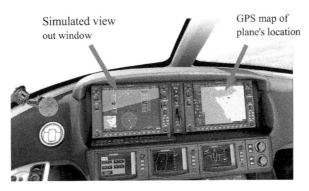

Figure 6.38. View from the cockpit of a simulated plane showing the synthetic vision view (left) and simulated GPS map of the plane's location (right). Created in X-Plane®.

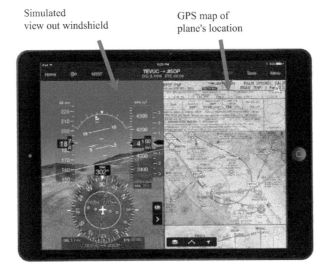

Figure 6.39. Synthetic vision display for a laptop. Copyright 2020 Garmin Ltd or its Subsidiaries. All Rights Reserved.

6.6 Other effects of light

6.6.1 Hyperbilirubinemia

Hyperbilirubinemia is a condition found in some premature newborns, 'premies'. Premies with this condition are born before their liver has developed fully. One of the liver's many functions is to remove from the bloodstream a substance called bilirubin. Bilirubin is formed when red blood cells break down. Too much bilirubin causes the premie's skin and sclera (whites of the eyes) to turn yellow—'hyperbilirubinemia'. An extreme accumulation of bilirubin is toxic. One effective treatment consists of directly exposing the premie to light with lots of blue in it. Proper eye protection is needed. The bilirubin absorbs the blue part of the applied light and breaks down into harmless components. This is an example of phototherapy. Figure 6.40 shows a color triangle; that the color yellow is opposite the color blue is no accident.

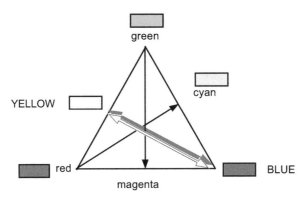

Figure 6.40. Color triangle showing the relationship between yellow and blue.

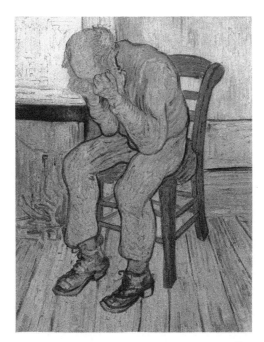

Figure 6.41. Vincent van Gogh, 'At Eternity's Gate.' Van Gogh was aware of the negative effects of a lack of light and spent much of his time in the south of France where the light levels are high. This image has been obtained by the author from the Wikimedia website https://commons.wikimedia.org/wiki/File:Van_Gogh_-_Trauernder_alter_Mann.jpeg, where it is stated to have been released into the public domain. It is included within this article on that basis.

6.6.2 Seasonal affective disorder

As seen above, exposure to light can affect us even though no images are formed. Some people react negatively to an overall reduction of light levels such as in winter. Called Seasonal Affective Disorder, SAD, it is also known as winter depression, or winter blues. This mood disorder is more prevalent the further north you go, most likely because winter light levels are lower in northern regions. Four out of five sufferers are women, and light therapy provides some relief, as does moving to Florida, Arizona, or the south of France, figure 6.41. Application of strong light helps in the therapeutic treatment of SAD, as well as depression, and chronic pain. Green light happens to be more effective than red, for unknown reasons. 'Green rooms' are used in theaters to relax performers before they go on stage. Other examples of non-imaging light affecting us is sunburn and the heating we feel from infrared rays when we stand in front of a roaring fire.

6.6.3 The parietal eye

Indirect light has other effects on us as well. Overall light levels affect the important operations of our pineal body and pituitary gland. The pineal body or parietal eye secretes melatonin. This substance is crucial for the regulation of our sleep cycles and our circadian rhythms (day–night cycles). A part of the pineal body is a group of photosensitive cells called 'the third eye'. In us, the third eye is buried deep within the brain and has no visual function. But in snakes, lizards and alligators, it is located at the surface; see figure 6.42, the third eye of an anole lizard.

The pineal gland secretes only one hormone (melatonin). In contrast, the pituitary gland secretes hormones that regulate blood pressure and metabolism.

Figure 6.42. The small gray circle in the highlighted rectangle is the third eye of the Carolina anole lizard (*Anolis carolinensis*). Credit: TheAlphaWolf. CC BY-SA 3.0.

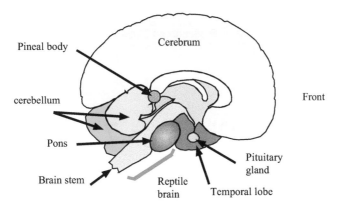

Figure 6.43. Cross-section of the human brain, showing the location of the pineal body ('parietal eye'), the pituitary gland, and the reptile brain.

For example, a friend of mine, not named Ralph, one day noticed that his shoes had shrunk, along with the rest of his clothes. Since it happened gradually over a period of time, the shrinkage went unobserved for a while. Once noticed, Ralph visited his physician, who found that Ralph's pituitary gland had a non-cancerous tumor. The tumor caused the pituitary to produce an excess of growth hormone, causing Ralph's body to grow in size. This gave the illusion that Ralph's clothes were shrinking. Once the tumor was removed, Ralph shrank back to his regular size and his shoes once again fit nicely.

Figure 6.43 shows some of the parts of the human brain, including the pineal body and the pituitary gland.

As an aside, we should take note of the area called the *reptile brain*, shown in figure 6.43 at the very bottom of the brain. Here we find the parts of the brain responsible for the most basic and fundamental functions of our brain. This region of our brain harks back to the dinosaurs and other reptiles such as alligators. Automatic survival actions such as threat response and fight or flight originate here. The reptile brain actions are immediate, non-logical, emotional, visual and non-auditory. Airplane pilots and police officers are trained to deliberately ignore the reptile brain's directions in emergency situations. Only then can the most logical and effective steps be taken in an emergency.

IOP Publishing

The Everyday Physics of Hearing and Vision (Second Edition)

Benjamin de Mayo

Chapter 7

Conclusions

And so now we come to the end of our journey through the worlds of seeing and hearing. I hope that you have gained a greater understanding and appreciation of your marvelous auditory and visual senses. Your hearing system allows you to distinguish 400 000 individual sounds with your 40 000 receptors per ear decoding each sound by frequency and intensity. Your visual system consists of 110 million rods and 40 million cones per eye. It enables you to see, with your unaided eyes, light from as far away as the Andromeda Galaxy, light that left there 2 billion years ago. This visual system resolves the tiniest of objects into complete images and enables you to see all of the myriad of colors and details present.

How fortunate we are indeed to possess such wonderful sensory systems as these!

Photo credit: James Wainscoat@tumbao1949 on Unsplash.

Lightning Source UK Ltd.
Milton Keynes UK
UKHW051640120921
390403UK00003B/40

9 780750 332057